Regression and Fitting on Manifold-valued Data

Ines Adouani · Chafik Samir

Regression and Fitting on Manifold-valued Data

Ines Adouani
Higher Institute of Applied Sciences
and Technology of Sousse (ISSAT)
University of Sousse
Sousse, Tunisia

Chafik Samir
University of Clermont Auvergne (UCA)
Aubière, France

ISBN 978-3-031-61711-9   ISBN 978-3-031-61712-6 (eBook)
https://doi.org/10.1007/978-3-031-61712-6

© The Editor(s) (if applicable) and The Author(s), under exclusive license to Springer Nature Switzerland AG 2024

This work is subject to copyright. All rights are solely and exclusively licensed by the Publisher, whether the whole or part of the material is concerned, specifically the rights of translation, reprinting, reuse of illustrations, recitation, broadcasting, reproduction on microfilms or in any other physical way, and transmission or information storage and retrieval, electronic adaptation, computer software, or by similar or dissimilar methodology now known or hereafter developed.
The use of general descriptive names, registered names, trademarks, service marks, etc. in this publication does not imply, even in the absence of a specific statement, that such names are exempt from the relevant protective laws and regulations and therefore free for general use.
The publisher, the authors and the editors are safe to assume that the advice and information in this book are believed to be true and accurate at the date of publication. Neither the publisher nor the authors or the editors give a warranty, expressed or implied, with respect to the material contained herein or for any errors or omissions that may have been made. The publisher remains neutral with regard to jurisdictional claims in published maps and institutional affiliations.

This Springer imprint is published by the registered company Springer Nature Switzerland AG
The registered company address is: Gewerbestrasse 11, 6330 Cham, Switzerland

If disposing of this product, please recycle the paper.

Contents

1 **Introduction** 1
  1.1 Motivation and Applications 1
  1.2 Challenges and Concerns 2
  1.3 Organization of This Book 5
  References 6

2 **Spline Interpolation and Fitting in $\mathbb{R}^n$** 9
  2.1 The Problem in a Continuous Setting 9
  2.2 Bézier Splines 11
    2.2.1 Bernstein Polynomials and Bézier Curves 11
    2.2.2 The de Casteljau Algorithm in $\mathbb{R}^n$ 15
    2.2.3 Bézier Splines and Continuity Conditions 15
  2.3 Solving Interpolation Problem in $\mathbb{R}^n$ 21
    2.3.1 C^1 Bézier Spline Interpolation in $\mathbb{R}^n$ 22
    2.3.2 C^2 Bézier Spline Interpolation in $\mathbb{R}^n$ 24
  References 26

3 **Spline Interpolation on the Sphere $\mathbb{S}^n$** 27
  3.1 Problem Statement 28
  3.2 Geometry of the Sphere $\mathbb{S}^n$ 30
  3.3 Spherical Bézier Splines 32
    3.3.1 Spherical Bézier Curves 32
    3.3.2 Spherical Bézier Splines and Continuity Conditions .. 35
  3.4 Solving Interpolation Problem on the Sphere $\mathbb{S}^n$ 38
    3.4.1 C^1 Spherical Bézier Spline 39
    3.4.2 C^2 Spherical Bézier Spline 43
  References 46

4 Spline Interpolation on the Special Orthogonal Group $SO(n)$ 49
- 4.1 Problem Formulation 49
- 4.2 Geometry of $SO(n)$ 51
- 4.3 Bézier Splines on $SO(n)$ 54
  - 4.3.1 Bézier Curves on $SO(n)$ 54
  - 4.3.2 Bézier Splines and Continuity Conditions 56
- 4.4 Solving Interpolation Problem on $SO(n)$ 57
  - 4.4.1 Proposed Interpolating Bézier Spline on $SO(n)$ 58
  - 4.4.2 Experiments 60
- References 62

5 Spline Interpolation on Stiefel and Grassmann Manifolds 65
- 5.1 Problem Formulation 65
- 5.2 Geometry of the Stiefel Manifold 67
- 5.3 Geometry of the Grassmann Manifold 69
- 5.4 Interpolation Problem on the Stiefel Manifold 71
  - 5.4.1 Bézier spline on the Stiefel Manifold 71
  - 5.4.2 C^1 Bézier Spline on the Stiefel Manifold 72
  - 5.4.3 Experiments 74
- 5.5 Interpolation Problem on the Grassmann Manifold 80
- References 82

6 Spline Interpolation on the Manifold of Probability Measures 85
- 6.1 Problem Formulation 85
- 6.2 Geometry of the Space of Probability Measures 86
  - 6.2.1 Manifold Structure 87
  - 6.2.2 Fisher–Rao Metric on $\mathcal{P}_+$ 90
  - 6.2.3 Geodesics on $\mathcal{P}_+$ 92
  - 6.2.4 Levi-Civita Parallel Transport on $\mathcal{P}_+$ 100
- 6.3 Interpolation Problem on Space of Probability Measures 105
  - 6.3.1 Measure Interpolation Spline Using De Casteljau Algorithm 106
  - 6.3.2 C^2 Bézier Splines on $\mathcal{P}_+$ 108
- 6.4 Experiments 109
  - 6.4.1 Numerical Examples 109
  - 6.4.2 Medical Examples 111
- References 112

7 Spline Interpolation on the Manifold of Probability Density Functions 115
- 7.1 Problem Formulation 116
- 7.2 Geometry of the Space of Probability Density Functions 117
  - 7.2.1 Fisher–Rao Geometry 117
  - 7.2.2 Riemannian Isometry 118
  - 7.2.3 Riemannian Structure of $\mathcal{M}$ 120
  - 7.2.4 Geodesic Curves on $\mathcal{P}$ 121

	7.3 Interpolation Problem on the Space of Probability Density Functions	123
	7.3.1 C^1 Spline on $\mathcal{M}$	123
	7.3.2 C^2 Bézier Splines on $\mathcal{M}$	124
	7.3.3 Properties of Splines on $\mathcal{P}$	127
7.4	Experiments	128
	References	135
8	**Spline Interpolation on Shape Space**	**137**
8.1	Problem Formulation	138
8.2	Geometry of Shape Space	139
	8.2.1 Curve Representation and Shape Space	139
	8.2.2 Geodesics in Shape Space, Exponential Map and Parallel Transport	140
8.3	Interpolation Problem on the Shape Space	141
8.4	Experiments	142
	8.4.1 Example 1	143
	8.4.2 Example 2	147
	References	148
9	**Spline Interpolation on Other Riemannian Manifolds**	**149**
9.1	Spline Interpolation on Symmetric Positive Definite Matrices $\mathcal{P}_n^+$	150
	9.1.1 Geometry of $\mathcal{P}_n^+$	150
	9.1.2 C^2 Bézier Spline on $\mathcal{P}_n^+$	151
9.2	Spline Interpolation on Hyperbolic Space $\mathcal{H}_n$	154
	References	155

Appendix A: Background Material 157

Index 179

Chapter 1
Introduction

This textbook addresses the challenges of fitting and interpolation on Riemannian manifolds. Initially, it delves into developing techniques for fitting data on a general Euclidean space. A significant emphasis will be placed on a specific aspect involving Bézier curves and the de Casteljau algorithm. Our particular attention is on Bézier spline interpolations due to their simplicity, flexibility, and theoretical guarantees. Subsequently, the exploration extends to the generalization of the proposed Euclidean Bézier curve techniques to various examples of Riemannian manifolds. Such generalization involves an in-depth examination of the geometric properties of the Riemannian manifold.

1.1 Motivation and Applications

Fitting smooth curves on non-linear spaces or manifolds is a fundamental task with diverse applications in machine learning, medical imaging, computer vision, and fields like Human Biometrics and Nanomanufacturing. In these applications, the common goal is to either estimate missing values, fill gaps in data, or model the dynamic behavior of objects. For instance, in medical imaging, data often represents anatomical shapes or structures on manifolds. For example, when evaluating tumor growth or decay post-medical diagnosis using medical images, employing shape analysis algorithms proves instrumental in identifying organ changes, facilitating the inference of disease progression [1]. Additionally, substantial interpolation occurs within the set of symmetric and positive-definite matrices(SPD) denoted as $\mathcal{P}_+^n$, particularly in tensor analysis [2, 3]. An illustrative example is seen in Diffusion Tensor Imaging (DTI) [4, 5], where spatial scans are characterized by 3D tensors, representing 3×3 symmetric and positive-definite matrices. The interpolation of DT-MRI data plays a pivotal role in deducing fiber architecture and gaining insights into specific diseases [6].

© The Author(s), under exclusive license to Springer Nature Switzerland AG 2024
I. Adouani and C. Samir, *Regression and Fitting on Manifold-valued Data*,
https://doi.org/10.1007/978-3-031-61712-6_1

Moreover, motion planning has been a prevalent and crucial subject in robotics since its inception. It finds applications in various domains such as robot rescue and robot service. Its objective is to play a significant role in controlling the smooth transition of a robot from one starting point to the next. Furthermore, the challenge of recovering structure and motion from a sequence of images, commonly referred to as stereo matching, remains a critical issue in computer vision. It continues to be one of the most actively researched areas, with remarkable advancements in imaging and computing hardware. This undertaking presents a particular challenge, especially in curve fitting within the special orthogonal group $SO(n)$ [7–9], the Stiefel manifold $St(n, p)$, and the Grassmann manifold $G_{n,p}(\mathbb{R})$ [10–12].

In the domain of statistics and data modeling, the process of fitting a set of Probability Density Functions (PDFs) is integral to numerous applications [13, 14]. Complex geometric data obtained from experiments, measurements, test methods, surveys, and more are often represented by PDFs to facilitate analysis, prediction, and classification. For example, consider a scenario where a video or at least two sets of images depict essentially the same scene but at different points in time. Objects in the images may undergo slight movement from one frame to another. Hence, predicting the motion direction of these objects or structures becomes desirable. PDFs can effectively model this motion, reducing the problem to a regression task within the space of PDFs. By fitting PDFs to the observed data, such as the spatial distribution of objects across frames, we can learn the underlying probability distribution governing the motion. This allows for the prediction of future movements or the detection of anomalous behavior based on deviations from the expected distribution. In this way, fitting PDFs provides a powerful framework for analyzing and understanding dynamic processes captured in data.

Beyond these applications, curve fitting also proves valuable in Human Biometrics. Various human biometric data, such as facial features, fingerprints, or body shapes, can be represented as points on Riemannian manifolds. The objective is to create accurate and smooth representations of biometric data, facilitating the prediction of variations in biometric features across different poses or expressions [15, 16].

In essence, the broad expanse of science and technology encompasses a myriad of applications, highlighting the importance of fitting non-vector data whether for prediction, estimating missing values, or recovering the motion of objects. Consequently, there is a growing need to explore innovative methods for accurately approximating smooth curves on non-linear manifolds.

1.2 Challenges and Concerns

Let $p_0, \ldots, p_N$ be a finite set of points on a Riemannian manifold $\mathcal{M}$, and let $0 = t_0 < \cdots < t_N = N$ represent a sequence of distinct and ordered instants of time. The challenge of fitting a smooth curve $\gamma : [t_0, t_N] \to \mathcal{M}$ to these specified points at the given times involves reconciling two competing objectives. Initially, the curve should

1.2 Challenges and Concerns

optimally align with the provided data. Simultaneously, it is necessary to guarantee that the curve maintains a certain level of "regularity." One approach to formalize this problem is to represent γ as the minimizer of an energy function defined over some suitable curve space Γ and given by

$$\min_{\gamma \in \Gamma} E(\gamma) = \frac{1}{2} \sum_{i=0}^{N} d_{\mathcal{M}}^2(\gamma(t_i), p_i) + \frac{\lambda}{2} \int_{t_0}^{t_N} \left< \frac{D^2\gamma(t)}{Dt^2}, \frac{D^2\gamma(t)}{Dt^2} \right> dt. \quad (1.1)$$

Here, $d_{\mathcal{M}}$ denotes the geodesic distance on $\mathcal{M}$, $\langle, \rangle$ represents the Riemannian metric, and $\frac{D^2\gamma}{Dt^2}$ stands for the second covariant derivative of γ. The parameter λ is used to fine-tune the trade-off between the competing objectives of fitting and smoothness.

When the Riemannian manifold $\mathcal{M}$ reduces to a Euclidean space, classical results suggest that the solution of (1.1) behaves like the interpolating natural cubic spline as $\lambda \to \infty$ and as the least-squares linear regression when $\lambda \to 0$. Nonetheless, extending classical methods for generating interpolating or smoothing splines from Euclidean spaces to manifolds is not as straightforward as one might expect. The primary challenge arises from the absence of explicit formulas for the analogs of polynomials in nonlinear spaces. An exception among polynomial functions in $\mathbb{R}^n$ is represented by straight lines, the natural generalization of which on Riemannian manifolds corresponds to geodesics. This challenge was initially addressed by Noakes et al. [17], who, similar to the solutions in Euclidean spaces [18], defined cubic polynomials on manifolds as curves minimizing the squared norm of the covariant acceleration. Subsequently, Camarinha et al. [19] introduced high-order Riemannian polynomials as solutions to the Euler-Lagrange equations associated with certain variational problems. However, due to the high nonlinearity of these differential equations, finding explicit solutions for Riemannian polynomials is extremely challenging, except for some trivial cases. For an in-depth exploration of significant theoretical contributions in this area, we refer to [20, 21].

Inspired by the definition of geometric polynomials on non-Euclidean spaces, Machado et al. [22] established a necessary condition for the optimality of (1.1) in the form of a fourth-order differential equation. Subsequently, they generalize the result and present an equivalent necessary condition for a solution to the problem (1.1) in its higher-order variational form [23].

The second challenge to highlight at the outset of solving the optimization problem (1.1) is that the curve space Γ over which E needs to be minimized to obtain the regression model has infinite dimension. As an attempt to overcome this issue, Samir et al. [24] choose to solve the optimization problem (1.1) through the application of a steepest descent scheme. They aim to minimize E on a suitable set of curves on the Riemannian manifold $\mathcal{M}$, equipped with the so-called Palais-metric in such a way that the gradient of E with respect to this metric is guaranteed to exist and be unique. Boumal et al. [25] suggest generating discrete representations of the optimal curve. This approach effectively reduces the curve space Γ to a finite-dimensional manifold, transforming the objective function (1.1) into an unconstrained, finite-dimensional optimization problem.

To address the challenges associated with the variational interpretation of (1.1), alternative approaches have been proposed. One such approach is grounded in Bézier curves, defined through a generalization to manifolds of the de Casteljau algorithm [18, 26]. In $\mathbb{R}^n$, this algorithm is a geometric construction joining two points with a polynomial through an iterative linear interpolation process. Its power lies in its geometric foundation, enabling easy generalization from $\mathbb{R}^n$ to other spaces, provided the linear interpolation process is suitably redefined. The algorithm has been extended for Riemannian manifolds [27–31], successfully addressing challenges encountered in the variational approach. This extension notably reduces the infinite-dimensional search space Γ to a finite-dimensional space composed of the control points of Bézier curves. This reduction significantly improves time and memory efficiency, and it also facilitates the enforcement of differentiability for the optimal curve, an essential feature in various applications. However, the de Casteljau algorithm functions as a computational tool only when explicit implementation details are worked out. Particularly, in contrast to Euclidean spaces, the challenge in generalizing de Casteljau lies in the difficulty of acquiring closed-form expressions for the minimal geodesic on manifolds.

In addition to constructing optimal interpolating Bézier curves as solutions to the optimization problem (1.1), the authors have also focused on addressing the problem of piecing these curves together into C^m splines, where $m \geq 0$. Bézier splines find applications across diverse fields such as computer graphics, animation, modeling, CAD, and CAGD. In numerous applications, there is a frequent need to either control the velocity and acceleration (potentially including higher-order derivatives) of an interpolating function or estimate these quantities from a sampled curve. Thus, the goal is to construct interpolants that exhibit relatively uniform speed and magnitude of acceleration, with a minimum requirement being C^2 continuity. Several works have introduced methods to construct interpolating Bézier splines on Riemannian manifolds of class C^1 [32–34]. Unfortunately, the problem of assembling Bézier curves into a C^2 spline is notably more complicated than achieving C^1 continuity, as it involves the computation of second derivatives. Nevertheless, in many practical applications, interpolants often require C^2 continuity, particularly in scenarios like rigid body motion planning, where abrupt changes in acceleration usually need to be avoided. Popiel et al. [29] was the first to successfully address the problem of piecing these curves together into C^2 splines for spherical data [29] and then generalize the result for symmetric Riemannian manifolds [30]. Equally important results have been developed in other works, for instance, [28, 31].

In this textbook, we endeavored to comprehensively address the challenges mentioned earlier. We introduce various numerical algorithms designed to construct approximating splines based on de Casteljau construction of Bézier curves. We specifically focus on Bézier spline interpolations due to their simplicity, flexibility, theoretical guarantees, and their property of minimizing mean acceleration. Initially, we formulate the energy minimization for linear least-squares based on Bézier curves in a general n-dimensional Euclidean space. Subsequently, we extend this concept to diverse instances of Riemannian manifolds $\mathcal{M}$. Notably, we establish conditions under which the Bézier spline on $\mathcal{M}$ achieves C^m continuity, $m = 0, 1, 2$. Such generalization involves an in-depth examination of the geometric properties of the Riemannian manifold, particularly its Riemannian metric, geodesics, Riemannian exponential map, and its inverse. These approaches prove adept at overcoming the difficulties associated with fitting data in Riemannian manifolds and boast several favorable properties. Numerical solutions not only exist but are also optimal in many common situations.

1.3 Organization of This Book

This textbook is organized as follows. Chapter 2 introduces the fitting and interpolation problem on the Euclidean space. We provide a review of the key concepts related to Bézier curves and Bézier splines, crucial for understanding the methods presented in this textbook. Additionally, we emphasize the prerequisites necessary to achieve C^m continuity, where $m = 0, 1, 2$. The chapter concludes by introducing a novel method to solve the interpolation problem in $\mathbb{R}^n$ using C^m Bézier splines.

We initiate the development of the proposed curve fitting approach in Riemannian manifolds by focusing on data representation on a sphere $\mathbb{S}^n$ of arbitrary dimension $n \geq 1$, in Chap. 3. This chapter introduces the generalization of the de Casteljau algorithm and Bézier curves to $\mathbb{S}^n$. We establish conditions under which the spherical Bézier spline on $\mathbb{S}^n$ achieves C^m continuity. Specifically, we delve into the computation of optimal intermediate control points that define a C^2 interpolating Bézier curve on $\mathbb{S}^n$.

We expand the methodology introduced in Chap. 3 to approximate C^2 Bézier splines on the compact Lie group $SO(n)$ in Chap. 4, the set of symmetric and positive-definite matrices $\mathcal{P}_+^n$, and hyperbolic spaces $\mathcal{H}_n$ characterized by constant negative curvature in Chap. 9. The geometry of each of these Riemannian manifolds is discussed in detail, and various numerical examples are provided in each case. These examples illustrate the necessity for smooth interpolating curves in these non-linear spaces for diverse real-world applications.

Shifting the attention to the search space of smooth regression splines, where the data points adhere to orthogonality constraints, we present an algorithm to approximate smooth curves on the Stiefel manifold $St(n, p)$ and the Grassmann manifold $G_{n,p}(\mathbb{R})$ in Chap. 5.

In Chap. 6, we explore the intricacies of fitting curves on the space of probability measures $\mathcal{P}_+$, endowed with the Fisher-Rao metric. We provide explicit theoretical expressions for essential geometric structures on $\mathcal{P}_+$, such as the Levi-Civita connection, minimal geodesics, parallel transport, exponential and logarithm maps. Building upon the method introduced in Chap. 3 for spherical data, we successfully apply these techniques to the space of probability measures.

Chapter 7 is dedicated to exploring methodologies for spline interpolations on the space of probability density functions (PDF) denoted as $\mathcal{P}$. $\mathcal{P}$ is a Riemannian manifold with infinite-dimensional characteristics, presenting a unique set of challenges when extending the regression methods discussed in previous chapters. To establish a connection, we leverage the Hilbert sphere equipped with the $\mathbb{L}^2$ metric, chosen for its advantageous properties and geometries in statistical analysis. Using the square-root representation of the probability density function, we proceed to construct a C^2 Bézier interpolating spline on $\mathcal{P}$.

Expanding on this context, the square-root representation of planar curves leads to a reduction of the set of curves equipped with the elastic metric. This transformation maps the space into a unit sphere within a Hilbert space equipped with the $\mathbb{L}^2$ metric, termed the shape space of curves. This, too, constitutes an infinite-dimensional Riemannian manifold.

Chapter 8 is dedicated to developing a framework for fitting a C^1 Bézier path to a given finite set of ordered data points within this latter Riemannian manifold.

Appendix A provides a concise overview of essential Riemannian geometry concepts. These concepts will play a crucial role throughout the book and are integral to comprehending the introduced curve fitting methodologies. This appendix is particularly helpful for readers who may not be familiar with these concepts, as it allows them to catch up and gain the necessary background understanding.

References

1. Grenander, U. and Miller, M.I. and Klassen, E. and Le, H. and Srivastava, A.: Computational anatomy: an emerging discipline, Quarterly of applied Mathematics, 4, 617-694, 1998.
2. Feragen, A. and Fuster, A.: Geometries and Interpolations for Symmetric Positive Definite Matrices, Modeling, Analysis, and Visualization of Anisotropy, Mathematics and Visualization (MATHVISUAL), 85-113, 2017.
3. Machado, L. and Silva Leite,F.: Approximating Sets of Symmetric and Positive-Definite Matrices by Geodesics, Conference Papers in Mathematics, 2013.
4. Westin, C-F. and Maier, E. and Mamata, H. and Nabavi, A. and Jolesz, F.A and Kikinis, R.: Process and Visualization for Diffusion Tensor MRI, Medical Image Analysis, 6, 93-108, 2002.
5. Jung, S. and Schwartzman, A. and Groisser, D.: Scaling-Rotation Distance and Interpolation of Symmetric Positive-Definite Matrices, SIAM Journal on Matrix Analysis and Applications, 36(3), 1180-1201, 2015.
6. Lazar, M. and Weinstein, D.M. and Tsuruda, J.S. and Hasan, K.M and Arfanakis, K. and Meyerand, M.E and Badie, B. and Rowley, H.A and Haughton, V.: White matter tractography using diffusion tensor deflection, Hum. Brain Mapp, 18(4), 306-321, 2003.
7. Zhao, X. and Liu, Y. and Xie, Z.: The application of acceleration-level quaternion interpolation in the visual-servoing process, 13th International Conference on Computer and Automation Engineering (ICCAE), 2021.
8. Legnani, G, and Fassi, I. and Tasora, A.:A practical algorithm for smooth interpolation between different angular positions Mechanism and Machine Theory, 2021.
9. Haarbach A. and Birdal, T. and Ilic, S.: Survey of higher order rigid body motion interpolation methods for keyframe animation and continuous-time trajectory estimation, International Conference on 3D Vision (3DV), 2018.
10. Man Lui,Y.: Advances in matrix manifolds for computer vision. ImageVis. Comput. 30(6-7), 380-388, 2012.
11. Zhang, D. and Balzano, L.:Global convergence of a Grassmannian gradient descent algorithm for subspace estimation. Proceedings of the 19th international conference on artificial intelligence and statistics, AISTATS, pp. 1460-1468. Cadiz, Spain, 2016.

References

12. Nguyen, T.S.: A real time procedure for affinely dependent parametric model order reduction using interpolation on Grassmann manifolds. Int. J. Numer. Meth. Eng. 93(8), 818-833, 2013.
13. Buckland, S.T.: Fitting Density Functions with Polynomials, Journal of the Royal Statistical Society, 41 (1), 63-76, 1992.
14. Clavijo-Blanco, J.A. and Gonzàlez Cagigal, M.A. and Rosendo-Macías, J.A.: A fitting procedure for probability density functions of service restoration times. Application to underground cables in medium-voltage networks, Electric Power Systems Research, 217, 2023.
15. Sidek, K.A, and Khalil, I: Enhancement of low sampling frequency recordings for ECG biometric matching using interpolation, Computer Methods and Programs in Biomedicine 109 (1), 13-25, 2013.
16. Bajahzar, A: Fingerprint Smoothing Using Different Interpolation Techniques, Journal of Engineering and Applied Sciences, 6 (2), 2019.
17. Noakes, L., Heinzinger,G. and Paden, B.: Cubic splines on curved spaces. IMA J. Math. Control Inform., 6(4):465-473, 1989.
18. Farin, G.E.: Curves and Surfaces for CAGD. Morgan Kaufmann, Academic Press, fifth edition, 2002.
19. Camarinha, M. and Silva Leite, F. and Crouch, P.:Splines of class C^k on Non-Euclidean Spaces, IMA Journal of Mathematical Control and Information, 12, 399-410, 1995.
20. Camarinha, M., Silva Leite, F. and Crouch. P.: On the Geometry of Riemannian Cubic Polynomials. Differential Geometry and its Applications 15, 107–135, 2001.
21. Giambo, R., Giannoni, F. and Piccione, P.: An Analytical Theory for Riemannian Cubic Polynomials. IMA Journal of Mathematical Control and Information, 19(4), December 2002.
22. Machado, L. and Silva Leite, F.: Fitting smooth paths on Riemannian manifolds, Int. J. Appl. Math. Stat., 4(J06):255-3, 2006.
23. Machado, L. and Silva Leite, F. and Krakowski, K.:High order smoothing splines versus least squares problems on Riemannian manifolds, Journal of Dynamical and Control Systems, 16(1), 121-148, 2010.
24. Samir, C. and Absil, P.-A. and Srivastava, A. and Klassen, E.: A Gradient-Descent Method for Curve Fitting on Riemannian Manifolds, Foundations of Computational Mathematics, 12, 49-73, 2012.
25. Boumal, N. and Absil, P.-A.: A discrete regression method on manifolds and its application to data on $SO(n)$, In IFAC Proceedings Volumes (IFAC-PapersOnline), 18, 2284-2289, 2011.
26. de Casteljau, P.: Outillages méthodes calcul. Technical Report André Citroen Automobiles, Paris 1959.
27. Park, F., and Ravani, B.: Bézier curves on Riemannian manifolds and Lie groups with kinematic applications, Trans. ASME J. Mech. Des. 117 36-40, 1995.
28. Crouch, P., Kun, G. and Silva Leite, F.: The de Casteljau algorithm on Lie groups and spheres, J. Dyn. Control Syst. 5, 397-429, 1999.
29. Popiel, T. and Noakes, L.: C^2 spherical Bézier splines, Comput. Aided Geom. Des. 23, 261-275, 2006.
30. Popiel, T. and Noakes, L.: Bézier curves and C^2 interpolation in Riemannian manifolds, J. Approx. Theory, 148(2), 111-127, 2007.
31. Geir, B. and Klas, M. and Olivier, V.:Numerical Algorithm for C^2-splines on Symmetric Spaces, SIAM Journal on Numerical Analysis, 56 (4), 2623-2647, 2018.
32. Gousenbourger, P.Y and Samir, C. and Absil, P.-A.: Piecewise-Bézier C1 interpolation on Riemannian manifolds with application to 2D shape morphing, In International Conference on Pattern Recognition (ICPR), 4086-4091, 2014.
33. Gousenbourger, P.Y., Massart, E. and Absil, P.A.: Data fitting on manifolds with composite Bézier-like curves and blended cubic splines, Journal of Mathematical Imaging and Vision, 61(5), 645-671, 2018.
34. Arnould, A., Gousenbourger, P.Y., Samir, C., Absil, P.A. and Canis, M.: Fitting Smooth Paths on Riemannian Manifolds: Endometrial Surface Reconstruction and Preoperative MRI-Based Navigation, Geometric Science of Information (GSI'15), 491-498, 2015.

Chapter 2
Spline Interpolation and Fitting in $\mathbb{R}^n$

This chapter unfolds a comprehensive exploration of the fitting and interpolation problem in $\mathbb{R}^n$. We present a formal definition of the fitting problem, simultaneously addressing the core challenge of accurately interpolating time-labeled data in Euclidean space. Subsequently, a thorough review of key definitions related to Bézier splines ensues, highlighting the prerequisites for achieving C^m continuity, ($m = 0, 1, 2$). The chapter culminates in the introduction of an innovative method for solving the interpolation problem in $\mathbb{R}^n$ through the use of C^m Bézier splines. This approach adeptly navigates the complexities of fitting data in multiple dimensions, ensuring the desired continuity up to the mth order and providing a nuanced and effective solution to this intricate problem.

2.1 The Problem in a Continuous Setting

Given $(N + 1)$ points $p_0, p_1, \ldots, p_N$ in $\mathbb{R}^n$ with time labels $0 = t_0 < t_2 < \cdots < t_N = 1$, the problem of curve fitting in the Euclidean space $\mathbb{R}^n$ is formulated as an optimization problem where the objective is to find a curve $\alpha : [t_0, t_N] \to \mathbb{R}^n$ that accurately represents the given points while exhibiting sufficient smoothness. Specifically, α is expressed as the curve that minimizes the weighted sum of two terms:

1. A sum-of-squares error penalizing the lack of fitting to the points,

$$E_d(\alpha) = \frac{1}{2} \sum_{i=0}^{N} \|p_i - \alpha(t_i)\|^2, \tag{2.1}$$

where $\|.\|$ denotes the usual 2-norm in $\mathbb{R}^n$.

© The Author(s), under exclusive license to Springer Nature Switzerland AG 2024
I. Adouani and C. Samir, *Regression and Fitting on Manifold-valued Data*,
https://doi.org/10.1007/978-3-031-61712-6_2

2. A regularity term E_s defined as follows:

- In the first case, it is expressed as the mean squared velocity of the curve,

$$E_{s,1}(\alpha) = \frac{1}{2}\int_{t_0}^{t_N} \|\dot{\alpha}(t)\|^2 dt. \tag{2.2}$$

- In the second case, it is expressed as the mean squared acceleration of the curve,

$$E_{s,2}(\alpha) = \frac{1}{2}\int_{t_0}^{t_N} \|\ddot{\alpha}(t)\|^2 dt. \tag{2.3}$$

The subsequent objective function, defined over the set of differentiable functions from $[t_0, t_N]$ to $\mathbb{R}^n$, $C^m([t_0, t_N], \mathbb{R}^n)$ for some appropriate $m \in \mathbb{N}$, encapsulates the crux of the problem:

$$E(\alpha) = E_d(\alpha) + \lambda E_{s,1}(\alpha) + \mu E_{s,2}(\alpha), \tag{2.4}$$

where tuning parameters λ and μ provide the flexibility to balance the trade-off between the fitting and smoothness objectives.

One possible approach for dealing with two objective functions is to transform the problem into a classical optimization setting. This entails designating one of the original objective function as the primary optimization goal, while the second objective function is recast as a constraint. The case of minimizing the energy function E under regularity constraints has been studied by different authors [1, 2]. The optimal solution is given in Theorem. 2.1.

Theorem 2.1 *If $p_0, p_2, \ldots, p_N$ are distinct points given in $\mathbb{R}^n$ with time labels $0 = t_0 < t_2 < \cdots < t_N = 1$, then there exists a unique polynomial α in the set $\mathbb{P}_r$ of polynomial functions with degree less than or equal to r, with $r \leq N$, that minimizes the functional $E_d(\alpha)$ defined by Eq. 2.1. The matrix X whose rows are the coefficients of the polynomial α is given by $X = (V^T V)^{-1} V^T P$, where matrices $V \in \mathbb{R}^{(r+1)\times(N+1)}$ and $P \in \mathbb{R}^{n\times(N+1)}$ are defined as follow*

$$V := \begin{pmatrix} 1 & t_0 & \cdots & t_0^r \\ 1 & t_1 & \cdots & t_1^r \\ \vdots & \vdots & & \vdots \\ 1 & t_N & \cdots & t_1^N \end{pmatrix} \quad \text{and} \quad P := \begin{pmatrix} q_0^1 & \cdots & t_0^n \\ q_1^1 & \cdots & q_1^n \\ \vdots & \vdots & \vdots \\ q_N^1 & \cdots & q_N^n \end{pmatrix}$$

Proof We refer the reader to [2] for the proof. □

Remark 2.1 The coefficients of the polynomial $\alpha \in \mathbb{P}_r$ that minimizes the energy functional E_d can be determined by solving a linear system of equations known as the "normal equations". Furthermore, when $r = N$, the polynomial curve α precisely interpolates the given data set of points at the specified instants of time.

2.2 Bézier Splines

In an alternative scenario, the optimization focuses on minimizing the regularity criterion E_s while adhering to a constraint on E_d, where the interpolation constraint $(\alpha(t_i) = p_i)$ is naturally imposed. It is well-established that when $\lambda > 0$ and $\mu = 0$, the optimal solution of the energy function (2.4) corresponds to a piecewise-linear interpolant. The progression from polynomial to piecewise polynomial functions forms the foundation of spline methods in curve fitting. These advanced techniques have demonstrated superior flexibility and efficacy compared to classical methods. Specifically, when $\lambda = 0$ and $\mu > 0$, the optimal solution of (2.4) under the same interpolation constraint gives rise to an approximating cubic spline. In this context, cubic splines are piecewise cubic polynomial curves of class C^2.

The curve space over which the objective function E must be minimized for the proposed regression model is infinite-dimensional. As discussed earlier, minimizing E in this context represents a intricate variational problem, particularly considering the primary focus in this textbook on solving fitting problems on manifolds rather than in $\mathbb{R}^n$. The inherent complexity of the continuous approach is a primary motivation for introducing Bézier splines in the next section. Instead of optimizing (2.4) over all functions in $C^m([t_0, t_N], \mathbb{R}^n)$, we aim to confine ourselves to a finite-dimensional space. A pragmatic approach involves considering a Bézier spline defined by composite Bézier curves. We can then optimize (2.4) over the space spanned by the control points of the spline. This strategy proves highly effective in the Euclidean case and extends very well to manifolds.

2.2 Bézier Splines

In this section, we first revisit some well-known definitions and formulas for Bernstein polynomials and Bézier curves. Bézier curves, named after Pierre Bézier who developed them in the 1960s, are widely used in computer graphics, CAD software, and curve fitting applications. Then, we define the Bézier spline and the conditions needed to obtain C^m-continuity, where $m = 0, 1, 2$, along this spline. For more details on Bézier curves and Bézier splines in Euclidean spaces, we refer to [3].

2.2.1 Bernstein Polynomials and Bézier Curves

Now, we delve into the definition of Bernstein polynomials, which are fundamental in the construction of Bézier curves.

Definition 2.1 Let $k \in \mathbb{N}$ and $0 \leq i \leq k$. The ith Bernstein polynomial of degree k is defined explicitly by

Fig. 2.1 Cubic Bernstein polynomials

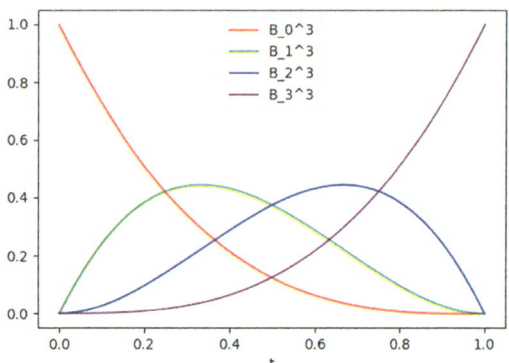

$$B_i^k(t) = \binom{k}{i} t^i (1-t)^{k-i}, 0 \leq t \leq 1, \quad (2.5)$$

where $\binom{k}{i} = \dfrac{k!}{i!(k-i)!}$ denotes the binomial coefficients.

Example 2.1 1. The Bernstein polynomials of degree 2, also called quadratic Bernstein polynomials, are given by

$$B_0^2(t) = (1-t)^2, \ B_1^2(t) = 2t(1-t)^2, \ B_2^2(t) = t^2, \ t \in [0,1].$$

2. The Bernstein polynomials of degree 3, also called cubic Bernstein polynomials, are given by (see Fig. 2.1)

$$B_0^3(t) = (1-t)^3, B_1^3(t) = 3t(1-t)^2, B_2^3(t) = 3t^2(1-t), B_3^3(t) = t^3, \ t \in [0,1].$$

Proposition 2.1 *Let $k \in \mathbb{N}$ and $0 \leq i \leq k$. The ith Bernstein polynomial of degree k possess the following properties:*

1. *They form a basis for the space of polynomials of degree less than or equal to k.*
2. *Recursion formula:*

$$B_i^k(t) = (1-t)B_i^{k-1}(t) + t B_{i-1}^{k-1}(t). \quad (2.6)$$

3. *Non-negativity:* $\forall t \in [0,1], \quad B_i^k(t) > 0.$
4. *Both sided end-point property:*

$$B_i^k(0) = \begin{cases} 1 & i=0 \\ 0 & i=1,...,k \end{cases} \quad \text{and} \quad B_i^k(1) = \begin{cases} 0 & i=0,...,k-1 \\ 1 & i=k \end{cases} \quad (2.7)$$

2.2 Bézier Splines

5. *Boundedness:*
$$0 \leq B_k^n(t) \leq 1, \quad 0 \leq t \leq 1. \tag{2.8}$$

6. *Partition of unity:*
$$\sum_{i=0}^{k} B_i^k(t) = \sum_{i=0}^{k-1} B_i^k(t) = 1. \tag{2.9}$$

7. *Symmetry:*
$$B_i^k(t) = B_{k-i}^k(t). \tag{2.10}$$

8. *Derivative formula:*
$$\frac{d}{dt} B_k^n(t) = n \left[B_{k-1}^{n-1}(t) - B_k^{n-1}(t) \right]. \tag{2.11}$$

Definition 2.2 Let $b_0, ..., b_k$ be $(k+1)$ points in $\mathbb{R}^n$. The Bézier curve $\beta_k : [0, 1] \to \mathbb{R}^n$ of order k is defined by

$$\beta_k(t; b_0, ..., b_k) = \sum_{i=0}^{k} b_i B_i^k(t), \tag{2.12}$$

where $B_i^k(t)$ denotes the ith Bernstein polynomial of degree k.

The Bézier curve β_k does not interpolate the points $b_0, ..., b_k$. Instead, these points are only used to generate the curve connecting b_0 and b_k, thereby influencing its shape. See Fig. 2.2.

Example 2.2 1. $\beta_1(t; b_0, b_1) = (1-t)b_0 + t b_1$ is a linear Bézier curve.
2. $\beta_2(t; b_0, b_1, b_2) = (1-t)^2 b_0 + 2(1-t)t b_1 + t^2 b_2$ is a Bézier curve of order 2, also known as a quadratic Bézier curve.
3. $\beta_3(t; b_0, b_1, b_2, b_3) = (1-t)^3 b_0 + 3t(1-t)^2 b_1 + 3t^2(1-t) b_2 + t^3 b_3$ is a Bézier curve of order 3, also known as a cubic Bézier curve.

Bézier curves exhibit various properties, many of which are inherited from the characteristics of Bernstein polynomials. For instance, the symmetry of Bernstein polynomials (2.10) implies the symmetry of Bézier curves: The order in which control points are labeled, whether in the ordinary or reverse order, doesn't affect the graph β_k. Equation 2.7 implies endpoint interpolation

$$\beta_k(0; b_0, ..., b_k) = b_0 \tag{2.13}$$
$$\beta_k(1; b_0, ..., b_k) = b_k \tag{2.14}$$

Equation 2.8 implies the convex hull property, indicating that the Bézier curve β_k of degree k lies inside the convex hull defined by its control points $b_0, ..., b_k$.

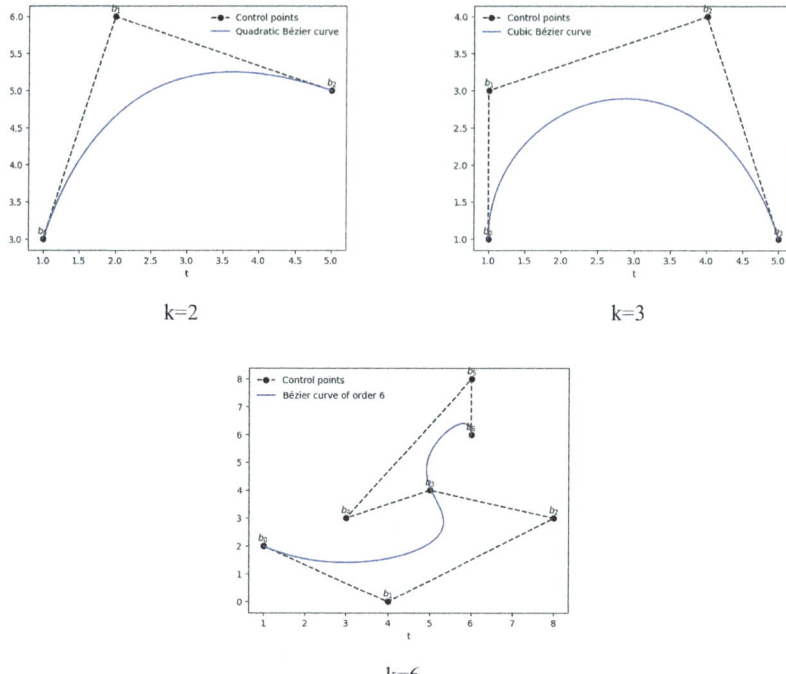

Fig. 2.2 Illustration of examples featuring a quadratic Bézier curve defined over the interval $[1, 5]$ with control points b_0, b_1 and b_2, a cubic Bézier curve defined over the same interval with control points b_0, b_1, b_2 and b_3 and a Bézier curve of order 6 defined over the interval $[1, 8]$ with control points b_0, b_1, b_2, b_3, b_4, b_5 and b_6

Equation 2.9 establishes the affine invariance of Bézier curves, meaning that the curve remains unchanged when control points undergo affine transformations. Using Eq. 2.11, one can determine the mth derivative of a Bézier curve:

$$\beta_k^m(t, b_0, ..., b_k) = \frac{n!}{(n-m)!} \sum_{k=0}^{n-m} \triangle^m b_k B_k^{n-m}(t) \qquad (2.15)$$

where $\triangle^m b_k = \triangle^{m-1} b_{k+1} - \triangle^{m-1} b_k$. Equation 2.15 then demonstrates the endpoint tangentiality in Bézier curves and also implies a nice criterion C^m-connection between two or more Bézier curves. Indeed, for $m = 1$, we have

$$\dot{\beta}_k(0; b_0, ..., b_k) = k(b_1 - b_0), \qquad (2.16)$$

$$\dot{\beta}_k(1; b_0, ..., b_k) = k(b_k - b_{k-1}). \qquad (2.17)$$

Finally, the simple recursion formula (2.6) leads to the well-known de Casteljau algorithm, which is employed for Bézier curve evaluation at a specific point $t \in [0, 1]$.

2.2 Bézier Splines

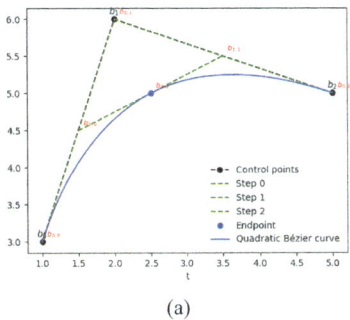
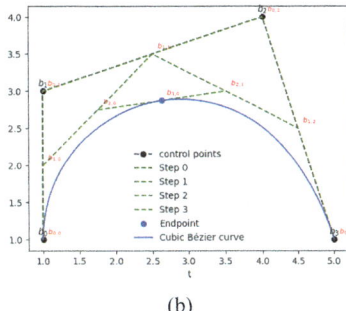

Fig. 2.3 Visualization of the de Casteljau algorithm illustrating its iterative steps and intermediate points for **a** a quadratic Bézier curve defined over the interval $[1, 5]$ and **b** a cubic Bézier curve defined over the same interval

2.2.2 The de Casteljau Algorithm in $\mathbb{R}^n$

The de Casteljau algorithm, introduced independently by de Casteljau [4] and Bézier [5], provides a fast and robust method for constructing the Bézier curve of degree k from the points b_i, $i = 0, .., k$, using successive linear interpolation techniques. The algorithm is recursive as follows: Given $b_0, ..., b_k$, $(k + 1)$ control points in $\mathbb{R}^n$ and $t \in [0, 1]$. Set $b_i^0 = b_0$, $i = 0, ..., k$, and then for each $j = 1, ..., k$, let

$$b_i^j = (1-t)b_i^{j-1} + tb_{i+1}^{j-1}, \quad i = 0, ..., k-j \text{ and } j = 1, ..., k. \tag{2.18}$$

The last point computed in this algorithm is the point on the Bézier curve β_k:

$$\beta_k(t; b_0, ..., b_k) = b_0^n. \tag{2.19}$$

Figure 2.3 illustrates the de Casteljau algorithm applied to the construction of a Bézier curve of order 2 and 3.

2.2.3 Bézier Splines and Continuity Conditions

Up to this point, the discussion on Bézier curves has predominantly focused on individual curves. Nevertheless, in various applications, there arises a requirement to assemble or blend segments from multiple curves to create a composite curve or a Bézier spline.

Definition 2.3 (*Piecewise Bézier Curve*) Let $0 = t_0 < t_1 < \cdots < t_N = 1$ be a distinct and ordered instants of time, and let $b_0^i, ..., b_k^i$ be a set of control points in $\mathbb{R}^n$, $i = 0, ..., N - 1$ and $k \in \mathbb{N}$. A piecewise Bézier curve, also referred to as a composite Bézier curve or a Bézier spline, formed by N Bézier curves of order k, is given by

$$\beta : [0, 1] \to \mathbb{R}^n, \quad t \to \beta_k^i(t - i, b_0^0, ..., b_k^i), t \in [i, i+1], i = 0, ..., N - 1. \tag{2.20}$$

Ensuring a certain degree of continuity between connected Bézier curves is desirable. Parametric continuity of order m, denoted as C^m, is achieved when the component functions of a parametric curve are m times differentiable with respect to the parameter within its specified interval. A piecewise Bézier curve, as defined above, is continuous if and only if $b_k^i = b_0^{i+1}$. This condition arises from the endpoint interpolation property for Bézier curves, ensuring the curve's single-valued nature. A piecewise Bézier curve that is continuous is said to have C^0 continuity. Higher orders of functional continuity are determined by the continuity of derivatives of the curve at the joint point $b_k^i = b_0^{i+1}$.

2.2.3.1 C^1 Continuity

For C^1 continuity, it is crucial that the ending slope of one Bézier curve matches the starting slope of the succeeding Bézier curve. This condition ensures that, in successive Bézier curves within the Bézier spline, the joining point between the curves is collinear with its adjacent control points. While it does ensure a tangentially continuous transition, the collinearity of three control points falls short of guaranteeing C^1 continuity. This is due to the nuanced interplay between the range and domain, which is essential for achieving C^1 continuity. In this textbook, the main focus is on cubic Bézier splines, which play a vital role in precisely defining and approximating curves using simple components. Consequently, the focus will be exclusively on presenting the conditions under which a cubic Bézier spline achieves C^1 continuity.

Proposition 2.2 *Consider a sequence β_3^i of N cubic Bézier curves, $i = 0, ..., N - 1$. The cubic Bézier spline defined by*

$$\beta(t) = \beta_3^i(t - i; b_0^0, ..., b_3^i), \quad \forall t \in [t_i, t_{i+1}] = [i, i+1], \quad i = 0, ..., N - 1, \tag{2.21}$$

achieves C^1 continuity, if the following condition holds

$$b_3^i = b_0^{i+1} = \frac{1}{2}(b_1^{i+1} + b_2^i), \quad i = 0, ..., N - 1. \tag{2.22}$$

2.2 Bézier Splines

Proof For $i = 0, ..., N - 1$, β_3^i are expressed explicitly in $\mathbb{R}^n$ with a number of control points $b_k^i, k = 0, ..., 3$ by

$$\beta_3^i(t; b_0^i, b_1^i, b_2^i, b_3^i) = (1-t)^3 b_0^i + 3t(1-t)^2 b_1^i + 3t^2(1-t)b_2^i + t^3 b_3^i. \quad (2.23)$$

It is easily seen that β is C^∞ on $]t_i, t_{i+1}[$, and is continuous if $b_3^i = b_0^{i+1}$, for $i = 0, ..., N - 1$. However, to ensure that β is C^1 at $t_i, i = 1, ..., N - 1$, we shall verify the following condition

$$\dot{\beta}_3^i(t - i + 1; b_0^i, b_1^i, b_2^i, b_3^i) = \dot{\beta}_3^{i+1}(t - i; b_0^i, b_1^i, b_2^i, b_3^i), \quad i = 0, ..., N - 1. \quad (2.24)$$

In fact, the first derivative of the Bézier curve β_3^i is:

$$\dot{\beta}_3(t; b_0^i, b_1^i, b_2^i, b_3^i) = -3(1-t)^2 b_0^i + 3(1 - 4t + 3t^2)b_1^i + 3t(2 - 3t)b_2^i + 3t^2 b_3^i. \quad (2.25)$$

Hence, the differentiability condition (2.24) gives

$$b_3^i - b_2^i = b_1^{i+1} - b_0^{i+1}, \quad i = 0, ..., N - 1, \quad (2.26)$$

or $b_3^i = b_0^{i+1}$. Thus we obtain

$$b_3^i = b_0^{i+1} = \frac{1}{2}(b_1^{i+1} + b_2^i), \quad i = 0, ..., N - 1. \quad (2.27)$$

which completes the proof. $\square$

Figure 2.4 depicts an example of a C^1 cubic Bézier spline defined by two cubic Bézier curves, and Algorithm 1 outlines the steps needed to construct a C^1 cubic Bézier spline.

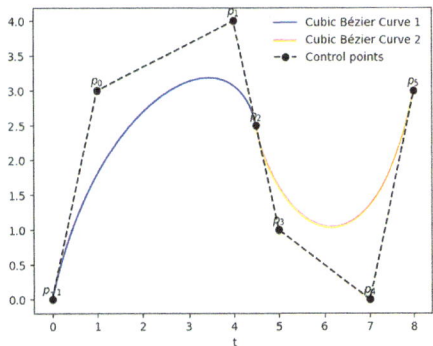

Fig. 2.4 C^1 cubic Bézier spline defined by connecting two cubic Bézier curves, illustrated with spline control points $d_{-1}, d_0, d_1, ..., d_5$

Algorithm 1 Construction of C^1 cubic Bézier spline

Input: $N \in \mathbb{N}$, $p_{-1}, p_0, p_1, ..., p_{2N}$ points in $\mathbb{R}^n$.
Output: C^1 Bézier spline β defined by control points $p_{-1}, p_0, p_1, ..., p_{2N}$.
1: For $i = 1, ..., N$, define the Bézier curves β_3^i as follow

$$\beta_3^i(t; b_{3i-3}, b_{3i-2}, b_{3i-1}, b_{3i})) = (1-t)^3 b_{3i-3} + 3t(1-t)^2 b_{3i-2} + 3t^2(1-t)b_{3i-1} + t^3 b_{3i}, i = 1, ..., N.$$

2: Set

$$b_0 = p_{-1}$$
$$b_1 = p_0$$
$$b_2 = p_1$$

3: For $i = 1, 2, ..., N-1$, compute the two central points for each intermediate Bézier curve β_3^i and calculate all points except the last one in the final Bézier curve β_3^N

$$b_{3i+1} = p_{2i} \quad \text{and} \quad b_{3i+2} = p_{2i+1}.$$

4: Define the last control point in the Bézier spline: $b_{3N} = p_{2N}$.
5: For $i = 1, 2, ..., N-1$, compute the joint points of Bezier curves:

$$b_{3i} = \frac{1}{2}(b_{3i-1} + b_{3i+1}).$$

6: **Return** $b_0, b_1, ..., b_{3N}$, the N Bézier curve β_3^i defined by control points $b_0, b_1, ..., b_{3N}$ and the C^1 cubic Bézier spline β.

Remark 2.2 1. The indexing for defining the spline control points starts at -1. This choice ensures that the last spline control point is given by $2N$, facilitating an easy determination of the number of Bézier curves used to define the Bézier spline.
2. As indicated by Algorithm 1, it is necessary to provide $(2N + 2)$ points to define a C^1 cubic Bézier spline with N cubic Bézier curves. This represents an advantage compared to the number of points required for creating a C^0 cubic Bézier spline, where $(3N + 1)$ points are needed (Fig. 2.5).

2.2.3.2 C^2 Continuity

In this section, we will establish the conditions under which a cubic Bézier spline, formed by a sequence of N cubic Bézier curves, achieves C^2 continuity. This level of continuity is a crucial property for curves, providing numerous advantages in diverse applications such as Computer-Aided Design (CAD), automotive design, computer graphics, and medical imaging. However, achieving C^2 continuity requires overcoming additional constraints.

2.2 Bézier Splines

Fig. 2.5 Visualization of C^1 cubic Bézier spline, defined by connecting two cubic Bézier curves. The spline is represented by control points $b_0, b_1, ..., b_6$. The illustration showcases the computation of control points for the two cubic Bézier curves and depicts iterative steps of the de Casteljau algorithm

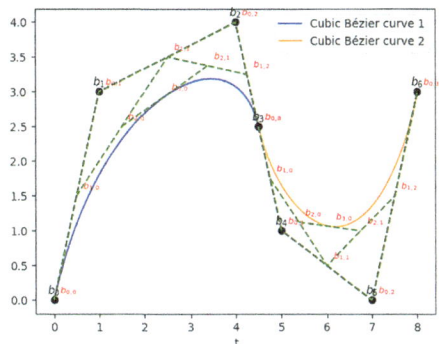

Proposition 2.3 *Let β_3^i be a sequence of N cubic Bézier curves defined by a set of control points b_3^i, $i = 0, ..., N - 1$. The cubic Bézier spline*

$$\beta(t) = \beta_3^i(t - i; b_0^0, ..., b_3^i), \quad \forall t \in [t_i, t_{i+1}] = [i, i+1], \quad i = 0, ..., N-1, \tag{2.28}$$

attains C^2 continuity, if the following conditions hold

$$b_3^i = b_0^{i+1} = \frac{1}{2}(b_1^{i+1} + b_2^i), \quad i = 0, ..., N-1, \tag{2.29}$$

$$b_1^i - 2b_2^i = b_2^{i+1} - 2b_1^{i+1}, \quad i = i = 0, ..., N-1. \tag{2.30}$$

Proof To ensure that β is C^2 at $t_i, i = 1, ..., N-1$, we shall make the following assumption

$$\ddot{\beta}_3^i(t - i + 1; b_0^i, b_1^i, b_2^i, b_3^i) = \ddot{\beta}_3^{i+1}(t - i; b_0^i, b_1^i, b_2^i, b_3^i), \quad i = 0, ..., N-1. \tag{2.31}$$

In fact, the second derivative of the Bézier curve β_3^i is:

$$\ddot{\beta}_3(t; b_0^i, b_1^i, b_2^i, b_3^i) = 6b_0^i - 12b_1^i + 6b_2^i + 6t(-b_0^i + 3b_1^i - 3b_2^i + b_3^i). \tag{2.32}$$

Hence, Eq. 2.31 gives

$$b_3^i - 2b_2^i + b_1^i = b_2^{i+1} - 2b_1^{i+1} + b_0^{i+1}, \quad i = i = 0, ..., N-1. \tag{2.33}$$

Considering the conditions of C^0 and C^1 continuity, Eq. 2.33 is modified as follows:

$$b_1^i - 2b_2^i = b_2^{i+1} - 2b_1^{i+1}, \quad i = i = 0, ..., N-1. \tag{2.34}$$

and the proof is complete. □

An illustration of a C^2 cubic Bézier spline, defined by two cubic Bézier curves, is presented in Fig. 2.7. Algorithm 2 summarizes the steps required to construct a C^2 cubic Bézier spline (Fig. 2.6).

Algorithm 2 Construction of C^2 cubic Bézier spline

Input: $N \in \mathbb{N}$, $p_{-2}, p_{-1}, p_0, p_1, ..., p_N$ points in $\mathbb{R}^n$.
Output: C^2 Bézier spline β defined by control points $p_{-2}, p_{-1}, p_0, p_1, ..., p_N$.
1: For $i = 1, ..., N$, define the Bézier curves β_3^i as follow

$$\beta_3^i(t; b_{3i-3}, b_{3i-2}, b_{3i-1}, b_{3i})) = (1-t)^3 b_{3i-3} + 3t(1-t)^2 b_{3i-2} + 3t^2(1-t) b_{3i-1} + t^3 b_{3i}, i = 1, ..., N$$

2: Set
$$b_0 = p_{-2}$$
$$b_1 = p_{-1}$$
$$b_{3N-1} = p_{N-1}$$
$$b_{3N} = p_N$$

3: Define
$$b_2 = \frac{1}{2} p_{-1} + \frac{1}{2} p_0 \text{ and } b_{3N-2} = \frac{1}{2} p_{N-2} + \frac{1}{2} p_{N-1}.$$

4: For $i = 1, 2, ..., N-2$, compute:
$$b_{3i+1} = \frac{2}{3} p_{i-1} + \frac{1}{3} p_i \text{ and } b_{3i+2} = \frac{1}{3} p_{i-1} + \frac{2}{3} p_i.$$

5: For $i = 1, 2, ..., N-1$, compute:
$$b_{3i} = \frac{1}{2}(b_{3i-1} + b_{3i+1}).$$

6: **Return** $b_0, b_1, ..., b_{3N}$, the N Bézier curve β_3^i defined by control points $b_0, b_1, ..., b_{3N}$ and the C^2 cubic Bézier spline β.

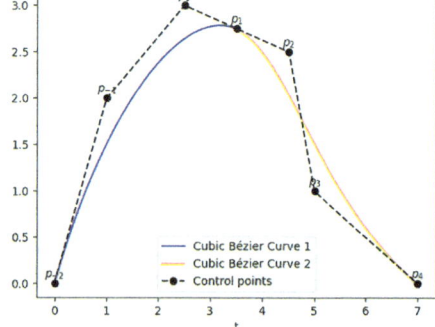

Fig. 2.6 A C^2 cubic Bézier spline, defined by connecting two cubic Bézier curves. The spline is represented by control points $p_{-2}, p_{-1}, p_0, p_1, p_2, p_3, p_4$

2.3 Solving Interpolation Problem in $\mathbb{R}^n$

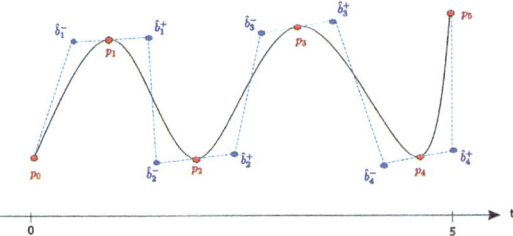

Fig. 2.7 Illustration of a composite cubic Bézier curve $\beta : [0, 5] \to \mathbb{R}$ composed of five individual Bézier curves: the curve joining points p_0 and p_1 is $\beta_2^0(t, p_0, \hat{b}_1^-, p_1)$, the curve joining points p_1 up to p_4 are cubic Bézier curve $\beta_3^i(t; p_{i-1}, \widehat{b}_{i-1}^+, \widehat{b}_i^-, p_i)$, $i = 1, ..., 3$. The curve joining points p_4 and p_5 is $\beta_2^4(t, p_4, \hat{b}_4^+, p_5)$

2.3 Solving Interpolation Problem in $\mathbb{R}^n$

Consider a sequence of $(N + 1)$ points in $\mathbb{R}^n$ denoted as $p_0, \ldots, p_N$, and let $t_0 < \ldots < t_n$ be a sequence of time instants. For simplicity, we assume that the time instants are given by $t_i = i$. In this section, the primary goal is to find a C^2 Bézier spline $\beta : [0, N] \to \mathbb{R}^n$ that interpolates the given set of points p_i: $\beta(t_i) = p_i$, where $i = 0, \ldots, N$, while minimizing the following cost function

$$\min_{\beta \in C^2([0,N],\mathbb{R}^n)} E(\beta) = \int_0^N \|\ddot{\beta}(t)\|^2 dt \quad (2.35)$$

In the construction of the Bézier spline β, we assume that it is defined by a sequence of N Bézier curves β_k^i, where $k \in \{2, 3\}$ and $i = 0, \ldots, N - 1$. Specifically, we consider the segments joining p_0 and p_1, as well as the segment joining p_{N-1} and p_N, to be Bézier curves of order two. All other segments are Bézier curves of order three. The Bézier curve β_k of degree $k \in \{2, 3\}$ is defined with a set of control points b_i, represented as their coefficients in the Bernstein basis polynomials. Explicitly, we have

$$\beta_2(t; b_0, b_1, b_2) = (1 - t)^2 b_0 + 2(1 - t)t b_1 + t^2 b_2,$$
$$\beta_3(t; b_0, b_1, b_2, b_3) = (1 - t)^3 b_0 + 3t(1 - t)^2 b_1 + 3t^2(1 - t)b_2 + t^3 b_3.$$

We introduce two artificial control points $(\widehat{b}_i^-, \widehat{b}_i^+)$ on the left and right-hand side of the interpolation point p_i for $i = 1, \ldots, (N - 1)$. Consequently, β is given by:

$$\beta(t) = \begin{cases} \beta_2(t; p_0, \widehat{b}_1^-, p_1), & \text{if } t \in [0, 1] \\ \beta_3(t - (i - 1); p_{i-1}, \widehat{b}_{i-1}^+, \widehat{b}_i^-, p_i), & \text{if } t \in [i - 1, i], i = 2, ..., N - 1 \\ \beta_2(t - (N - 1); p_{N-1}, \widehat{b}_{N-1}^+, p_N), & \text{if } t \in [N - 1, N] \end{cases}$$

It is easy to check that the interpolation conditions $\beta_i(t_i)$ are satisfied, and $\beta|_{[t_i,t_{i+1}]}$ is smooth. The current objective is to seek the control points $\widehat{b_i^-}$ and $\widehat{b_i^+}$ that generate the Bézier curve β, a solution to the optimization problem (2.35).

2.3.1 C^1 Bézier Spline Interpolation in $\mathbb{R}^n$

In order to attain C^1 continuity, it is necessary to fulfill the differentiability condition at the joint points, as indicated in Proposition 2.2. In other words, we shall make the following assumption:

$$\dot{\beta}_{k^i}(b_0^i, ..., b_{k^i}^i; t - i + 1)|_{t=i} = \dot{\beta}_{k^{i+1}}(b_0^{i+1}, ..., b_{k^{i+1}}^{i+1}; t - i)|_{t=i} \quad i = 0, ..., N - 2. \tag{2.36}$$

This differentiability condition allows us to express $\widehat{b_i^+}$ in terms of $\widehat{b_i^-}$ as:

$$\widehat{b_1^+} = \frac{5}{3}p_1 - \frac{2}{3}\widehat{b_1^-}, \tag{2.37}$$

$$\widehat{b_i^+} = 2p_i - \widehat{b_i^-}, i = 2, ..., N - 2, \tag{2.38}$$

$$\widehat{b_{N-1}^+} = \frac{5}{2}p_{N-1} - \frac{3}{2}\widehat{b_{N-1}^-}. \tag{2.39}$$

Consequently the optimization problem (2.35) can be formulated as follows:

$$\min_{\widehat{b_1^-},...,\widehat{b_{N-1}^-}} E(\widehat{b_1^-}, ..., \widehat{b_{N-1}^-}) := \min_{\widehat{b_1^-},...,\widehat{b_{N-1}^-}} \int_0^1 \|\ddot{\beta}_2^0(t; p_0, \widehat{b_1^-}, p_1)\|^2 +$$

$$\sum_{i=1}^{N-2} \int_0^1 \|\ddot{\beta}_3^i(t; p_i, \widehat{b_i^+}, \widehat{b_{i+1}^-}, p_{i+1})\|^2 + \int_0^1 \|\ddot{\beta}_2^{N-1}(t; p_{N-1}, \widehat{b_{N-1}^-}, p_N)\|^2 \tag{2.40}$$

Theorem 2.2 *If $p_0, ..., p_N$ are distinct points given in $\mathbb{R}^n$ and $t_0 < \cdots < t_N$ is a given partition of the unit time interval $[0, N]$, then there exists a unique $X \in R^{N-1 \times n}$ that minimizes the cost function (2.40). The rows of the matrix X are control points of the Bézier curve β.*

Proof Minimizing the functional E requires the computation of its gradient and then the search of its critical points. As the energy is based on polynomials, it is possible to compute the analytical expression of its gradient. We first compute the acceleration on respective intervals and then evaluate the integral of each term of Eq. 2.40. For brevity, we skip the details and we give just the final formulation of Eq. 2.40, which allows us to determine the optimal solution. After simplification, we have:

2.3 Solving Interpolation Problem in $\mathbb{R}^n$

$$\min_{\widehat{b}_1^-,\dots,\widehat{b}_{N-1}^-} E(\widehat{b}_1^-,\dots,\widehat{b}_{N-1}^-) := 4(-4p_0^T\widehat{b}_1^- + 4(\widehat{b}_1^-)^T\widehat{b}_1^- - 4p_1^T\widehat{b}_1^-)$$

$$+ \sum_{i=1}^{N-2} 36(-p_i^T\widehat{b}_i^+ + (\widehat{b}_i^+)^T\widehat{b}_i^+ - (\widehat{b}_i^+)^T\widehat{b}_{i+1}^-$$

$$+ (\widehat{b}_{i+1}^-)^T\widehat{b}_{i+1}^- - (\widehat{b}_{i+1}^-)^T p_{i+1})$$

$$+ 4(-4p_{N-1}^T\widehat{b}_{N-1}^+ + 4(\widehat{b}_{N-1}^+)^T\widehat{b}_{N-1}^+ - 4p_N^T\widehat{b}_{N-1}^+)$$

$$+ K \tag{2.41}$$

Where K denotes a constant term involving all p_i. Next we replace $\widehat{b}_i^+$'s by $\widehat{b}_i^-$'s using Eq. 2.38. Then the optimal solution is given by:

$$\frac{\partial E}{\partial \widehat{b}_1^-} = 4(8\widehat{b}_1^- - 4p_0 - 4p_1) + 36(-\frac{14}{9}p_1 + \frac{8}{9}\widehat{b}_1^- + \frac{6}{9}\widehat{b}_2^-) = 0,.$$

$$\frac{\partial E}{\partial \widehat{b}_2^-} = 36(-\frac{5}{3}p_1 + \frac{2}{3}\widehat{b}_1^- + 2\widehat{b}_2^- - p_2) + 36(-3p_2 + 2\widehat{b}_2^- + \widehat{b}_3^-) = 0. \tag{2.42}$$

$$\frac{\partial E}{\partial \widehat{b}_j^-} = 36(-3p_j + 2\widehat{b}_j^- + \widehat{b}_{j+1}^-) + 36(-2p_{j-1} + \widehat{b}_{j-1}^- + 2\widehat{b}_j^- - p_j) = 0 \quad j = 2,\dots,N-2.$$

$$\frac{\partial E}{\partial \widehat{b}_{N-1}^-} = 36(-2p_{N-2} + \widehat{b}_{N-2}^- + 2\widehat{b}_{N-1}^- - p_{N-1}) + 4(-24p_{N-1} + 18\widehat{b}_{N-1}^- + 6p_N) = 0.$$

We exploit the fact that the sum of the coefficients in each equation is equal to zero and express all equations as a linear system of the form $AX = CP$, with X being the matrix of variables $\widehat{b}_i^-$ of size $(N-1) \times n$, A a matrix of size $(N-1) \times (N-1)$, C a matrix of size $(N-1) \times (N+1)$ and P the matrix of p_i's of size $(N+1) \times n$:

$$A := \begin{pmatrix} 16 & 6 & & & & & \\ 6 & 36 & 9 & & & & \\ & 9 & 36 & 9 & & & \\ & & \ddots & \ddots & \ddots & & \\ & & & 9 & 36 & 9 & \\ & & & & \ddots & \ddots & \ddots \\ & & & & & 9 & 36 & 9 \\ & & & & & & 9 & 36 \end{pmatrix}, \quad X := \begin{pmatrix} \widehat{b}_1^- \\ \vdots \\ \widehat{b}_{N-1}^- \end{pmatrix},$$

$$C := \begin{pmatrix} 4 & 18 & & & & & \\ & 15 & 36 & & & & \\ & & 18 & 36 & & & \\ & & & \ddots & \ddots & & \\ & & & & 18 & 36 & \\ & & & & & \ddots & \ddots \\ & & & & & 18 & 36 \\ & & & & & & 18 & 33 & -6 \end{pmatrix} \qquad P := \begin{pmatrix} p_0 \\ \vdots \\ p_N \end{pmatrix}.$$

Where A is a tridiagonal sparse square matrix with a dominant diagonal leading to a unique optimal solution:

$$X = A^{-1}CP = DP \text{ with } \sum_{j=0}^{N} D_{ij} = 1, \forall i. \tag{2.43}$$

This concludes the proof. Detailed steps are summarized in Algorithm 3. □

Algorithm 3 Construction of interpolating C^1 Bézier spline in $\mathbb{R}^n$

Input: $N \geq 3, n \geq 2, P = [p_0, ..., p_N]^T$ a matrix of size $n \times (N + 1)$ containing the $(N + 1)$ interpolation points.
Output: $X = [\widehat{b}_1^-, ..., \widehat{b}_{N-1}^-]^T$
1: Define the functional E with (2.40).
2: Compute the gradient of E.
3: Compute the critical points of the gradient of E.
4: Form linear system with matrix equation $AX = CP$ where A is a matrix of size $(n - 1) \times (n + 1)$ and C is a matrix of size $(N - 1) \times (N + 1)$ given by Eqs. (2.37)–(2.39).
5: Solve the linear system.
6: **Return** X,

2.3.2 C^2 Bézier Spline Interpolation in $\mathbb{R}^n$

Now, let us assume that β is C^1, ensuring that (2.36) is satisfied, and the solution X is obtained. The additional C^2 condition for a C^1 spline is the equality of the second derivative at the joint point p_i as explained in Proposition 2.3. In fact, in this case, we have,

$$\ddot{\beta}_{k^i}(b_0^i, ..., b_{k^i}^i; t - i + 1)|_{t=i} = \ddot{\beta}_{k^{i+1}}(b_0^{i+1}, ..., b_{k^{i+1}}^{i+1}; t - i)|_{t=i} \quad i = 0, ..., N - 2. \tag{2.44}$$

2.3 Solving Interpolation Problem in $\mathbb{R}^n$

Clearly, with this C^2 condition, the positions of the control point $\widehat{b}_i^-$ and $\widehat{b}_i^+$ that generate the curve β will be modified. Therefore, it is more convenient to introduce a different notation. Let b_i^- and b_i^+ denote the new control points on the left and right-hand side of the interpolation point p_i. By computing the acceleration of β on the respective intervals and considering Eqs. 2.37–2.39, we obtain:

$$b_2^- = \frac{1}{3}p_0 - \frac{1}{2}b_1^- + \frac{8}{3}p_1, \tag{2.45}$$

$$b_{i+1}^- = b_{i-1}^+ + 4p_i - 4b_i^-, \, i = 2, \ldots, N-2, \tag{2.46}$$

$$p_N = 2p_{N-1} + 2b_{N-1}^+ - 6b_{N-1}^- + 3b_{N-2}^+. \tag{2.47}$$

The points affected by the additional C^2 condition are $\widehat{b}_i^-$ and, consequently, $\widehat{b}_i^+$, for $i = 2, \ldots, N-1$. The point $\widehat{b}_1^-$ remains invariant, and similarly $\widehat{b}_1^+$ remains unchanged. Thus, $b_1^- = \widehat{b}_1^-$, where $\widehat{b}_1^-$ represents the first row of the matrix X obtained as a solution to the optimization problem (2.35). However, the endpoint p_N is affected as deduced from Eq. 2.47. Nevertheless, providing the control point b_1^- enables us to determine all the other control points, including b_2^- using Eq. 2.45 and hence b_2^+ with Eq. 2.38. Subsequently, b_{i+1}^- for $i = 2, \ldots, N-2$ can be found with Eq. 2.46 and consequently, b_i^+, for $i = 3, \ldots, N-2$ with Eq. 2.38 and b_{N-1}^+ with Eq. 2.39.

Algorithm 4 details the procedure for constructing an interpolating C^2 Bézier spline in $\mathbb{R}^n$.

Algorithm 4 Construction of interpolating C^2 Bézier spline in $\mathbb{R}^n$

Input: $N \geq 3, n \geq 2, P = [p_0, \ldots, p_N]^T$ a matrix of size $n \times (N+1)$ containing the $(N+1)$ interpolation points.
Output: $Y, \widehat{P}$
1: Determine $X = [\widehat{b}_1^-, \ldots, \widehat{b}_{N-1}^-]^T$ using Algorithm 3.
2: Set : $b_1^- = \widehat{b}_1^-$
3: Calculate control point $b_1^+ : b_1^+ = \frac{5}{3}p_1 - b_1^-$.
4: Calculate control point $b_2^- : b_2^- = \frac{1}{3}p_0 - \frac{1}{2}b_1^- + \frac{8}{3}p_1$.
5: **for** $i = 2 : N-2$ **do**
6: $\quad b_i^+ = 2p_i - b_i^-$
7: $\quad b_{i+1}^- = b_{i-1}^+ + 4p_i - 4b_i^-$
8: **end for**
9: Calculate control point $b_{N-1}^+ : b_{N-1}^+ = \frac{5}{2}p_{N-1} - \frac{3}{2}b_{N-1}^-$.
10: Determine $\widehat{P} = [\widehat{p}_0, \ldots, \widehat{p}_N]^T$ a matrix of size $n \times (N+1)$ containing the new $(N+1)$ interpolation points:
11: **for** $i = 0 : N-1$ **do**
12: $\quad \widehat{p}_i = p_i$
13: **end for**
14: Compute interpolation point $\widehat{p}_N: \widehat{p}_N = 2p_{N-1} + 2b_{N-1}^+ - 6b_{N-1}^- + 3b_{N-2}^+$.
15: **Return** $Y, \widehat{P}$,

Remark 2.3 An alternative derivation of the C^2 condition introduces the existence of additional control points d_i for $i = 0, ..., N-2$, such that:

$$b_1^- = \frac{3}{5}p_0 + \frac{2}{5}d_0 \tag{2.48}$$

$$b_1^+ = \frac{3}{5}d_0 + \frac{2}{5}b_2^- \tag{2.49}$$

Then,

$$b_i^- = \frac{1}{2}b_{i-1}^+ + \frac{1}{2}d_{i-1}, \quad i = 2, ..., N-2 \tag{2.50}$$

$$b_i^+ = \frac{1}{2}d_{i-1} + \frac{1}{2}b_{i+1}^-, \quad i = 2, ..., N-2 \tag{2.51}$$

Finally,

$$b_{N-1}^- = \frac{3}{5}b_{n-2}^+ + \frac{2}{5}d_{n-2} \tag{2.52}$$

$$b_{N-1}^+ = \frac{3}{5}d_{n-2} + \frac{2}{5}p_n \tag{2.53}$$

Therefore, to construct a C^2 Bèzier spline β with the property $\beta(t_i) = p_i$, we start by selecting, as the first row of the matrix b_1^- as a first row of the matrix X, a solution to 2.36. Subsequently, we determine d_0 using Eq. 2.48 and b_1^+ with Eq. 2.37. We then calculate b_2^- with Eq. 2.49. This process is iterated to obtain d_i, b_i^- and b_i^+ using Eqs. 2.50–2.53. It is important to note that both methods for verifying C^2 continuity are mathematically equivalent.

References

1. Lancaster, P., Salkauskas K.: Curve and Surface Fitting. Academic Press, 1990.
2. Machado, L. and Silva Leite, F.: Fitting smooth paths on Riemannian manifolds, Int. J. Appl. Math. Stat., 4(J06):255-3, 2006.
3. Farin, G.E.: Curves and Surfaces for CAGD. Morgan Kaufmann, Academic Press, fifth edition, 2002.
4. de Casteljau, P.: Outillages méthodes calcul. Technical Report André Citroen Automobiles, Paris 1959.
5. Bezier, P. The mathematical basis of the UNISURF CAD System. Butterworths, London, 1986.

Chapter 3
Spline Interpolation on the Sphere $\mathbb{S}^n$

Spherical splines have applications across various domains, including cardiology, computer vision, geophysics, biology, astronomy, animation, robotics, and motion planning utilizing quaternions [1–4]. One noteworthy example occurs in vector cardiograms, where the electrical activity of the heart during a heartbeat is represented as a nearly planar orbit in $\mathbb{R}^3$. In this context, spherical splines offer an effective approach for modeling and analyzing the intricate dynamics of cardiac electrical signals. These splines provide a flexible and efficient framework to capture the complexities of the heart's electrical behavior, facilitating the interpretation and diagnosis of cardiac conditions [1, 2]. Another significant application arises in computer graphics and animation, specifically with spherical splines on the unit sphere $\mathbb{S}^3$. This utilization enables the smooth representation of orientations of solid bodies, as quaternions can be interpreted as pairs of antipodal points on $\mathbb{S}^3$. Spherical spline curves, in this case, offer a means to specify a smooth transitions of solid orientations, contributing to the creation of visually appealing and realistic animations in computer graphics [3–5].

In this chapter, we address the problem of approximating smooth splines on the sphere $\mathbb{S}^n$ of arbitrary dimension $n \geq 1$. The theoretical foundation behind fitting and interpolating points on the unit sphere $\mathbb{S}^n$ is presented, along with a review of the generalization of the de Casteljau algorithm and Bézier curves to $\mathbb{S}^n$. Then, we establish conditions under which the Bézier spline on $\mathbb{S}^n$ achieves C^m continuity, where $m = 0, 1, 2$. Finally, the proposed approach to compute optimal intermediate control points defining a C^2 interpolating Bézier curve on $\mathbb{S}^n$ is introduced and proven to adeptly address the challenges of fitting data in Riemannian manifolds while enjoying several favorable properties.

3.1 Problem Statement

Let $x_0, \ldots, x_N$ be a finite set of $(N + 1)$ data points on $\mathbb{S}^n$, $n \geq 2$, and let $t_0 < \cdots < t_N$ be a sequence of distinct and ordered instants of time. For simplicity, we assume that the time instants are given by $t_i = i$. The continuous objective function (2.4) has been generalized to $\mathbb{S}^n$ to take this form:

$$E(\gamma) = \frac{1}{2} \sum_{i=0}^{N} d^2(\gamma(t_i), p_i) + \frac{\lambda}{2} \int_{t_0}^{t_N} <\dot{\gamma}(t), \dot{\gamma}(t)> dt \qquad (3.1)$$
$$+ \frac{\mu}{2} \int_{t_0}^{t_N} <\frac{D^2\gamma(t)}{Dt^2}, \frac{D^2\gamma(t)}{Dt^2}> dt,$$

where d denotes the geodesic distance on $\mathbb{S}^n$, $<,>$ the Riemannian metric, $\dot{\gamma}$ the first derivative of γ and $\frac{D^2\gamma}{Dt^2}$ the second covariant derivative of γ. λ, $\mu \in \mathbb{R}_+$ are two parameters used to fine-tune the trade-off between the competing objectives of fitting and smoothness. Similarly to $\mathbb{R}^n$, the previous equation can be defined as follows:

$$E(\gamma) = E_d(\gamma) + \lambda E_{s,1}(\gamma) + \mu E_{s,2}(\gamma). \qquad (3.2)$$

An alternative approach to address an optimization problem with two objective functions involves optimizing a weighted sum of these functions. In the context of curve fitting on a general Riemannian manifold $\mathcal{M}$, Machado et al. [6] pursued this approach using the first-order smoothing term $E_{s,1}(\gamma)$, while authors in [7] focused on the second-order smoothing term $E_{s,2}(\gamma)$. More precisely, in [6], the objective function is formulated as:

$$E(\gamma) = \frac{1}{2} \sum_{i=0}^{N} d^2(\gamma(t_i), p_i) + \frac{\lambda}{2} \int_{t_0}^{t_N} <\dot{\gamma}(t), \dot{\gamma}(t)> dt, \qquad (3.3)$$

over the class of all piecewise smooth curves $\gamma : [0, 1] \to \mathcal{M}$. Solutions to this variational problem on $\mathcal{M}$ yield piecewise smooth geodesics that effectively capture the given data [6, Theorem 3.8]. As demonstrated in [6, Theorem 3.9], as the parameter λ goes to $+\infty$, the optimal curve converges to a singular point. Notably, this singular point is identified as the Riemannian mean of the data points on $\mathcal{M}$. Conversely, when λ tends to zero, the optimal curve tends towards a broken geodesic on $\mathcal{M}$ that interpolates the provided data points.

On the other hand, in [7], the objective function is formulated as:

$$E(\gamma) = \frac{1}{2} \sum_{i=0}^{N} d^2(\gamma(t_i), p_i) + \frac{\lambda}{2} \int_{t_0}^{t_N} <\frac{D^2\gamma(t)}{Dt^2}, \frac{D^2\gamma(t)}{Dt^2}> dt, \qquad (3.4)$$

3.1 Problem Statement

over a certain set of admissible C^2 curves. The authors establish a necessary condition of optimality, which takes the form of a fourth-order differential equation involving the covariant derivative and the curvature tensor, along with certain regularity conditions at the time instants $t_i, i = 0, ..., N$, [7, Theorem 4.4]. The optimal curves are approximating cubic splines, they are considered approximating because, in general, $\gamma(t_i)$ differs from (p_i), and they are cubic splines because they are obtained by smoothly piecing together segments of cubic polynomials on $\mathcal{M}$, following the approach of Noakes et al. [8]. Furthermore, it is demonstrated in [7, Proposition 4.5] that as the smoothing parameter λ goes to $+\infty$, the optimal curve converges to the best least-squares geodesic fit to the data points at the given instants of time. Conversely, when λ approaches zero, the approximating cubic spline converges to an interpolating cubic spline [7, Proposition 4.6].

Unfortunately, the variational approach, apparently designed to generalize polynomials on Riemannian manifolds, encounters the challenge of solving highly nonlinear differential equations. Consequently, obtaining an explicit solution for the optimal curve on Riemannian manifolds becomes an exceedingly difficult task. In this endeavor, numerous methods have been proposed to approximate data on the sphere $\mathbb{S}^n$. In [9], Machado et al. introduced a numerical optimization algorithm, rooted in the variational procedure, to tackle the geodesic regression problem $\sum_{i=1}^{i=N} d^2(p_i, \gamma(t_i))$ on the sphere $\mathbb{S}^n$. This problem is a natural extension of the classical regression problem 2.1 in the Euclidean case but proves to be more intricate due to the higher nonlinearity of the normal equations. To tackle this challenge, the authors reformulated the geodesic regression problem on the sphere as a constrained nonlinear optimization problem in the embedding Euclidean space $\mathbb{R}^{n+1}$ and employed numerical techniques for its solution. In [10], they also provided a numerical computation of the solution to the spline variational problem (3.4) on $\mathbb{S}^n$. Additionally, Jupp and Kent [11] proposed a regression method on spherical smoothing splines using rolling and unwrapping techniques. This approach inspired different researchers to address interpolating and approximating smoothing splines problems in other different manifolds [12, 17].

An alternative method for approximating spherical data involves extending the approach of Bézier curves and the de Casteljau algorithm. This extension entails replacing line segments with the shortest circular arcs. Specifically, [5] generalized the de Casteljau algorithm to a three-dimensional unit sphere and provided a condition for a uniform C^1 spline. Simultaneously, in [13], the application of the de Casteljau algorithm was explored on compact Lie groups and spheres. In [14], the focus was on defining conditions, expressed in terms of control points, for smoothly connecting Bézier curves into C^2 Bézier splines on the sphere $\mathbb{S}^n$. Recently, Kwan-Young et al. [15] introduced an intrinsic curve fitting technique that relies on the generalized Bézier curve and incorporates a local penalization scheme on $\mathbb{S}^2$. This method aims to regularize the disparities in velocity vectors between consecutive linear Bézier curves used to construct the higher-order Bézier spline.

In this chapter, we aim to find $\gamma : [0, N] \to \mathbb{S}^n$, a solution to (3.4), taking the form of a Bézier spline and satisfying the following properties: (i) $\gamma(t_i) = x_i$ (ii) γ is of class C^2. To achieve this, we reduce the interpolation problem to the classical

Euclidean setting, allowing us to directly leverage the extensive toolbox of spline interpolation. Inspired by [14], we establish conditions under which the interpolating Bézier spline on $\mathbb{S}^n$ attains C^2 continuity.

3.2 Geometry of the Sphere $\mathbb{S}^n$

Let $\mathbb{S}^n = \{x \in \mathbb{R}^{n+1}, \ \|x\| = 1\}$ be the unit sphere in $\mathbb{R}^{n+1}$. This is an embedded submanifold of $\mathbb{R}^{n+1}$.

Lemma 3.1 *The tangent space to $\mathbb{S}^n$ at a point $x \in \mathbb{R}^{n+1}$ is the vector subspace of $\mathbb{R}^{n+1}$ defined by*

$$T_x\mathbb{S}^n = \{v \in \mathbb{R}^{n+1}, \ x^t v = 0\}. \tag{3.5}$$

Proof Consider the function $f : \mathbb{R}^{n+1} \to \mathbb{R}$ given by $f(x) = x^t x - 1$. The sphere $\mathbb{S}^n$ can be expressed as the set $\{x \in \mathbb{R}^{n+1}, \ f(x) = 0\}$. Since f is a smooth function and is full rank on $\mathbb{R}^{n+1}$, the differential of f is $Df(x) = 2x^t v$ and the tangent space is the kernel of $Df(x)$, which leads to $T_x\mathbb{S}^n = \ker(Df(x)) = \{v \in \mathbb{R}^{n+1}, \ x^t v = 0\}$. □

The Euclidean inner product on the tangent vectors makes $\mathbb{S}^n$ a Riemannian manifold. For any $v_1, v_2 \in T_x\mathbb{S}^n$, the Riemannian metric is used as $<v_1, v_2> = v_1^t v_2$. A point $x \in \mathbb{S}^n$ and a vector $v \in T_x\mathbb{S}^n$ uniquely determine a geodesic $t \to \alpha(t)$ under the Euclidean metric. This geodesic passes through x at $t = 0$ with velocity v and is given by:

$$\alpha(t; x, v) = \cos(\|v\|t)x + \frac{v}{\|v\|}\sin(\|v\|t). \tag{3.6}$$

Lemma 3.2 *Let x and y two non-antipodal points on $\mathbb{S}^n$. Then for all $t \in [0, 1]$, The shortest geodesic that joins x to y, is represented by:*

$$\alpha(t; x, y) = \cos(\theta t) + \frac{\sin(\theta t)}{\sin(\theta)}(y - x\cos(\theta)) \tag{3.7}$$

$$= \frac{\sin((1-t)\theta)}{\sin(\theta)}x + \frac{\sin(t\theta)}{\sin(\theta)}y, \tag{3.8}$$

where $\theta = \cos^{-1}(<x, y>) \in]0, \pi[$ represents the geodesic distance between two points on $\mathbb{S}^n$. The function α satisfies the following properties:

1. $\alpha(0, x, y) = x$ and $\alpha(1, x, y) = y$.
2. $\alpha(t, x, y) = \alpha(1 - t, y, x)$.

Consequently, the exponential map $\exp_x : T_x\mathbb{S}^n \to \mathbb{S}^n$ has a simple expression

$$\exp_x(v) = \cos(\|v\|)x + \frac{v}{\|v\|}\sin(\|v\|). \tag{3.9}$$

3.2 Geometry of the Sphere $\mathbb{S}^n$

The exponential map is a bijection when we limit $\|v\|$ to the interval $[0, \pi)$, and we have $\exp_x(v) = -y$ for all v with $\|v\| = \pi$. Thus, for a point $y \in \mathbb{S}^n$ ($x \neq y$), the inverse exponential map $\exp_x^{-1}(y)$ (also known as the logarithmic map Log) is given by

$$\exp_x^{-1}(y) = \frac{\theta}{\sin(\theta)}(y - x\cos(\theta)). \tag{3.10}$$

Henceforth, the derivative of α is expressed as $\dot\alpha(t) = \frac{\partial}{\partial u}|_{u=t}\alpha(t)$, and the derivative of $\exp_x$ at $v \in T_x\mathbb{S}^n$ will be denoted as $(d\exp_x)_v$ and computed by

$$(d\exp_x)_v(w) = \frac{d}{dt}|_{t=0} \exp_x(v + tw), \quad \text{for all } w \in T_x\mathbb{S}^n \simeq T_v(T_x\mathbb{S}^n). \tag{3.11}$$

The parallel transport of a vector v along a geodesic curve α joining $\alpha(0) = x$ and $\alpha(1) = y$ is the linear isometry $\Gamma_{\alpha(0)\to\alpha(1)} : T_x\mathbb{S}^n \to T_y\mathbb{S}^n$ given by

$$\Gamma_{x\to y}(v) = v - 2\frac{x+y}{\|x+y\|^2}<v, y>. \tag{3.12}$$

The symmetry of the Riemannian manifold $\mathbb{S}^n$ is an intriguing property and is crucial for demonstrating the main results of the following section. Specifically, for any $x \in \mathbb{S}^n$, the geodesic symmetry that reverses geodesics through x is represented by,

$$\varphi_x(y) = -y + 2<y, x>x, \tag{3.13}$$

In the ensuing discussion, $(d\phi_x)_y$ denotes the derivative of ϕ_x at y.

Lemma 3.3 *The global symmetry $\varphi_x : \mathbb{S}^n \to \mathbb{S}^n$ has the following properties:*

1. $(d\varphi_x)_y^{-1} = (d\varphi_x)_{\varphi_x(y)}$, *for all $y \in \mathbb{S}^n$.*
2. $(d\varphi_x)_{Exp_x(v)} \circ (dExp_x)_v = -(dExp_x)_{-v}$, *for all $v \in T_x\mathbb{S}^n$.*

The complete set of symmetries constitutes the entire isometry group of $\mathbb{S}^n$, identified as the orthogonal group $O(n+1)$. The Lie group $O(n+1)$ exhibits a smooth action on the unit sphere $\mathbb{S}^n$. For any point $x \in \mathbb{S}^n$, the isotropy subgroup is conjugate to $O(n)$. Consequently, $\mathbb{S}^n$ stands as a symmetric homogeneous space, and it is diffeomorphic to the quotient manifold $O(n+1)/O(n)$.

The geometric toolbox described in Table 3.1 summarizes closed-form expressions of geometric structures on the sphere $\mathbb{S}^n$.

Table 3.1 Geometric toolbox for the Sphere $\mathbb{S}^n$

Set	$\mathbb{S}^n = \{x \in \mathbb{R}^{n+1},\ \|x\| = 1\}$
Tangent spaces	$T_x\mathbb{S}^n = \{v \in \mathbb{R}^{n+1},\ x^t v = 0\}$
Inner product	$<v_1, v_2> = v_1^T v_2$
Geodesic distance	$d(x, y) = \theta = \cos^{-1}(x, y)$
Shortest geodesic	$\alpha(t, x, y) = \frac{\sin((1-t)\theta)}{\sin(\theta)} x + \frac{\sin(t\theta)}{\sin(\theta)} y$
Exponential	$\exp_x(v) = \cos(\|v\|)x + \frac{v}{\|v\|}\sin(\|v\|)$
Logarithm	$\text{Log}_x(y) = \frac{\theta}{\sin(\theta)}(y - x\cos(\theta))$
Parallel transport	$\Gamma_{x \to y}(v) = v - 2\frac{x+y}{\|x+y\|^2} <v, y>$
Geodesic symmetry	$\varphi_x(y) = -y + 2 <y, x> x$

3.3 Spherical Bézier Splines

We now give a method for constructing smooth splines on the sphere. To address this, we introduce spherical Bézier splines and outline their fundamental properties. Additionally, we discuss the conditions under which these splines achieve C^m continuity, considering the cases for $m = 0, 1, 2$.

3.3.1 Spherical Bézier Curves

3.3.1.1 The de Casteljau Algorithm on $\mathbb{S}^n$

Given $(k + 1)$ points $x_0, x_1, ..., x_k$ on $\mathbb{S}^n$, where the geodesic distance between consecutive points is less than π, and $t \in [0, 1]$. The spherical Bézier curve $t \to \beta_k(t; x_0, ..., x_k)$ of order k on the sphere $\mathbb{S}^n$ defined with control points $x_0, ..., x_k$, can be generated as follows: Initially, set $x_i^0 = x_i$, $i = 0, ..., k$. Then, for $i = 0, ..., k - j$ and $j = 1, ..., k$, compute the points x_i^j using the recursive formula:

$$x_i^j = \alpha\left(t, x_i^{j-1}, x_{i+1}^{j-1}\right), \quad t \in [0, 1],$$

where $\alpha(t, x_i^{j-1}, x_{i+1}^{j-1})$ denote the geodesic curve joining x_i^{j-1} and x_{i+1}^{j-1} according to Eq. 3.7. The last point computed in this algorithm represents a point on the Bézier curve β_k:

$$x_0^k = \beta_k(t; x_0, ..., x_k). \tag{3.14}$$

Example 3.1 The cubic Bézier curve $\beta_3(t; x_0, x_1, x_2, x_3)$ on the sphere $\mathbb{S}^n$, defined by the control points x_0, x_1, x_2, x_3, can be constructed using the de Casteljau algo-

3.3 Spherical Bézier Splines

rithm as follows. Fix $t \in [0, 1]$, for $i = 0, 1, 2, 3$, set $x_0^0 = x_0$, $x_1^0 = x_1$, $x_2^0 = x_2$ and $x_3^0 = x_3$. We remind that

$$\alpha(t, x_i^{j-1}, x_{i+1}^{j-1}) = \frac{\sin((1-t)\theta_i^j)}{\sin(\theta_i^j)} x_i^{j-1} + \frac{\sin(t\theta_i^j)}{\sin(\theta_i^j)} x_{i+1}^{j-1}, \quad (3.15)$$

where $\theta_i^j = \cos^{-1}(<x_i^{j-1}, x_{i+1}^{j-1}>) \in]0, \pi[$. Now, for $i = 0$ and $j = 1$, we compute points x_0^1, x_1^1, x_2^1 as follows:

$$x_0^1 = \alpha(t, x_0^0, x_1^0) = \frac{\sin((1-t)\theta_0^1)}{\sin(\theta_0^1)} x_0^0 + \frac{\sin(t\theta_0^1)}{\sin(\theta_0^1)} x_1^0, \quad \theta_0^1 = \cos^{-1}(<x_0, x_1>)$$

$$x_1^1 = \alpha(t, x_1^0, x_2^0) = \frac{\sin((1-t)\theta_1^1)}{\sin(\theta_1^1)} x_1^0 + \frac{\sin(t\theta_1^1)}{\sin(\theta_1^1)} x_2^0, \quad \theta_1^1 = \cos^{-1}(<x_1, x_2>)$$

$$x_2^1 = \alpha(t, x_2^0, x_3^0) = \frac{\sin((1-t)\theta_2^1)}{\sin(\theta_2^1)} x_2^0 + \frac{\sin(t\theta_2^1)}{\sin(\theta_2^1)} x_3^0, \quad \theta_2^1 = \cos^{-1}(<x_2, x_3>)$$

Now, when i ranges over 0 and 1, and $j = 2$, we compute points x_0^2 and x_1^2 as follows:

$$x_0^2 = \alpha(t, x_0^1, x_1^1) = \frac{\sin((1-t)\theta_0^2)}{\sin(\theta_0^2)} x_0^1 + \frac{\sin(t\theta_0^2)}{\sin(\theta_0^2)} x_1^1, \quad \theta_0^2 = \cos^{-1}(<x_0^1, x_1^1>)$$

$$x_1^2 = \alpha(t, x_1^1, x_2^1) = \frac{\sin((1-t)\theta_1^2)}{\sin(\theta_1^2)} x_1^1 + \frac{\sin(t\theta_1^2)}{\sin(\theta_1^2)} x_2^1, \quad \theta_1^2 = \cos^{-1}(<x_1^1, x_2^1>)$$

Finally, the point

$$x_0^3 = \alpha(t, x_0^2, x_1^2) = \frac{\sin((1-t)\theta_0^3)}{\sin(\theta_0^3)} x_0^2 + \frac{\sin(t\theta_0^3)}{\sin(\theta_0^3)} x_1^2, \quad \theta_0^3 = \cos^{-1}(\langle x_0^2, x_1^2 \rangle)$$

is the point on the cubic Bézier curve β_3 for the fixed $t \in [0, 1]$ (Fig. 3.1).

3.3.1.2 Properties

Proposition 3.1 *The Bézier curve $t \to \beta_k(t; x_0, ..., x_k)$ of order k satisfies:*

1. $\beta_k(0; x_0, ..., x_k) = x_0$ and $\beta_k(1; x_0, ..., x_k) = x_k$.
2. $\beta_k(t; x_0, ..., x_k) = \beta_k(1-t; x_k, ..., x_0), t \in [0, 1]$.

Proof The proof follows easily from Lemma 3.2. □

Let's denote $\dot{\beta}_k(t; x_0, ..., x_k) = \frac{d}{du}\big|_{u=t} \beta_k(u; x_0, ..., x_k)$ and $\frac{D}{du}\big|_{u=t} \beta_k(u; x_0, ..., x_k)$ as the velocity and covariant acceleration of the Bézier curve β_k of order k, where $t \in [0, 1]$.

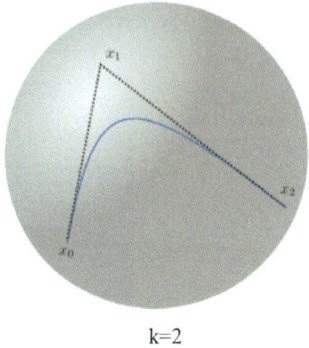
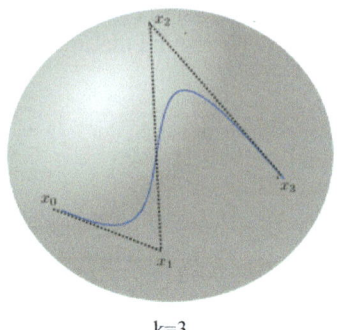

Fig. 3.1 Illustration of examples featuring on the left a Bézier curve of degree 2 defined on $\mathbb{S}^2$ with control points x_0, x_1, and x_2, and on the right a Bézier curve of degree 3 defined on $\mathbb{S}^2$ with control points x_0, x_1, x_2, and x_3

Proposition 3.2 *The derivatives of the Bézier curve $t \to \beta_k(t; x_0, ..., x_k)$ of order k satisfy the following boundary conditions:*

1. $\dot{\beta}_k(0; x_0, ..., x_k) = k\dot{\alpha}(0, x_0, x_1) = k \log_{x_0}(x_1)$.
2. $\dot{\beta}_k(1; x_0, ..., x_k) = k\dot{\alpha}(1, x_{k-1}, x_k) = k \log_{x_k}(x_{k-1})$.

Proof We refer the reader to [14] for the proof. □

Example 3.2 The derivatives of a cubic Bézier curve $\beta_3(t; x_0, x_1, x_2, x_3)$ on $\mathbb{S}^n$ satisfy

$$\dot{\beta}_3(0; x_0, x_1, x_2, x_3) = 3\dot{\alpha}(0, x_0, x_1) = 3\left(-\theta_0^1 \cot(\theta_0^1) x_0 + \frac{\theta_0^1}{\sin(\theta_0^1)} x_1\right), \quad (3.16)$$

$$\dot{\beta}_3(0; x_0, x_1, x_2, x_3) = 3\dot{\alpha}(0, x_2, x_3) = 3\left(-\theta_2^1 \cot(\theta_2^1) x_3 + \frac{\theta_2^1}{\sin(\theta_2^1)} x_2\right), \quad (3.17)$$

where $\theta_0^1 = \cos^{-1}(<x_0, x_1>)$ and $\theta_2^1 = \cos^{-1}(<x_2, x_3>)$.

Proposition 3.1 underscores the endpoint interpolation property for the spherical Bézier curve and establishes its symmetry on $\mathbb{S}^n$. Similarly, Proposition 3.2 reveals that the first (and respectively, last) derivative of the curve aligns with the geodesic connecting the two initial (and respectively, two final) control points. Consequently, it becomes evident that all the properties observed in Bézier curves within Euclidean space $\mathbb{R}^n$ are retained for the spherical Bézier curve on $\mathbb{S}^n$. Moving forward, we will now proceed to compute the endpoint covariant accelerations of a spherical Bézier curve.

3.3 Spherical Bézier Splines

Theorem 3.1 *The covariant acceleration of the Bézier curve $t \to \beta_k(t; x_0, ..., x_k)$ of order k satisfies*

1. $\frac{D}{dt}|_{t=0} \dot{\beta}_k(t; x_0, ..., x_k) = k(k-1)\Omega_0$, where

$$\Omega_0 := \begin{cases} \dot{\alpha}(0, x_1, x_2), & if\, x_0 = x_1 \\ \frac{1}{\sin(\theta_0^1)h(x_0, x_1, x_2)}, & if\, x_0 \neq x_1 \end{cases}$$

and
$h(x_0, x_1, x_2) = \theta_0^1 \dot{\alpha}(0, x_1, x_2) - \sin(\theta_0^1)\dot{\alpha}(0, x_0, x_1)$
$+ \langle \dot{\alpha}(0, x_1, x_2), x_0 \rangle \left(\frac{\dot{\alpha}(1, x_0, x_1)}{\sin(\theta_0^1)} - \frac{\dot{\alpha}(0, x_0, x_1)}{\theta_0^1} \right)$.

2. $\frac{D}{dt}|_{t=1} \dot{\beta}_k(t; x_0, ..., x_k) = k(k-1)\Omega_k$, where

$$\Omega_k := \begin{cases} -\dot{\alpha}(0, x_{k-2}, x_{k-1}), & if\, x_{k-1} = x_k \\ \frac{1}{\sin(\theta_{k-1}^1)h(x_{k-2}, x_{k-1}, x_k)}, & if\, x_{k-1} \neq x_k \end{cases}$$

and
$h(x_{k-2}, x_{k-1}, x_k) = -\theta_{k-1}^1 \dot{\alpha}(1, x_{k-2}, x_{k-1}) - \sin(\theta_{k-1}^1)\dot{\alpha}(1, x_{k-1}, x_k) +$
$\langle \dot{\alpha}(1, x_{k-2}, x_{k-1}), x_k \rangle \left(\frac{\dot{\alpha}(0, x_{k-1}, x_k)}{\sin(\theta_0^{k-1})} - \frac{\dot{\alpha}(1, x_{k-1}, x_k)}{\theta_{k-1}^1} \right)$.

Proof See [14] for the proof. □

3.3.2 Spherical Bézier Splines and Continuity Conditions

Definition 3.1 Let $0 = t_0 < t_1 < ... < t_N = 1$ be distinct and ordered instants of time, and let $x_0^i, ..., x_k^i$ be a set of control points on $\mathbb{S}^n$, $i = 0, .., N-1$ and $k \in \mathbb{N}$. A Spherical Bézier spline on $\mathbb{S}^n$, formed by N spherical Bézier curves of order k^i, is defined by

$$\beta : [0, N] \to \mathbb{S}^n, \quad t \to \beta_{k^i}^i(t - i, x_0^0, ..., x_{k^i}^i), \quad t \in [i, i+1], \quad i = 0, ..., N-1. \tag{3.18}$$

The spherical Bézier spline, as defined above is C^∞ on $]t_i, t_{i+1}[$ and is continuous if and only if $x_{k^i}^i = x_0^{i+1}$. This condition arises from the endpoint interpolation property for spherical Bézier curves. A spherical Bézier spline that is continuous is said to have C^0 continuity. To achieve C^1 continuity, it is essential to satisfy the differentiability condition at the junction points, $x_{k^i}^i = x_0^{i+1}$.

Proposition 3.3 *The spherical Bézier spline $\beta : [0, N] \to \mathbb{S}^n$ defined by (3.18) is C^1 if the following conditions hold*

$$x_{k^i}^i = x_0^{i+1} \tag{3.19}$$

$$k^i \dot{\alpha}(1, x_{k^i-1}^i, x_{k^i}^i) = k^{i+1} \dot{\alpha}(0, x_0^{i+1}, x_1^{i+1}) \tag{3.20}$$

Proof Suppose that β is continuous, the spherical Bézier spline β achieves C^1 continuity if and only if

$$\dot{\beta}_{k^i}^i(t - i + 1; x_0^i, ..., x_{k^i}^i)|_{t=i} = \dot{\beta}_{k^{i+1}}^{i+1}(t - i; x_0^{i+1}, ..., x_{k^{i+1}}^{i+1})|_{t=i} \quad i = 0, ..., N-1. \tag{3.21}$$

Applying Proposition 3.2, we show that Eq. 3.21 holds if and only if

$$k^i \dot{\alpha}(1, x_{k^i-1}^i, x_{k^i}^i) = k^{i+1} \dot{\alpha}(0, x_0^{i+1}, x_1^{i+1}), \tag{3.22}$$

which completes the proof. $\square$

Figures 3.1, 3.2 illustrates an example of a C^1 Cubic Bézier spline with 3 cubic Bézier curves on $\mathbb{S}^2$.

Example 3.3 Consider a spherical cubic Bézier spline formed by two cubic spherical Bézier curves β_3^i on $\mathbb{S}^n$. The spherical cubic Bézier spline is defined as

$$\beta(t) = \begin{cases} \beta_3(t; x_0^0, x_1^0, x_2^0, x_3^0), & \text{if } t \in [0, 1] \\ \beta_3(t - 1; x_0^1, x_1^1, x_2^1, x_3^1), & \text{if } t \in [1, 2] \end{cases}$$

Let's assume that β is continuous, this gives $x_3^0 = x_0^1$. Then, β is C^1 if and only if $\dot{\alpha}(1, x_2^0, x_3^0) = \dot{\alpha}(0, x_0^1, x_1^1)$. From (3.16) and (3.17), we deduce that

$$\left(-\theta_0^1 \cot(\theta_0^1) x_0^1 + \frac{\theta_0^1}{\sin(\theta_0^1)} x_1^1\right) = \left(\frac{\theta_2^1}{\sin(\theta_2^1)} x_2^0 - \theta_2^1 \cot(\theta_2^1) x_0^1\right), \tag{3.23}$$

where $\theta_0^1 = \cos^{-1}(< x_0^1, x_1^1 >)$ and $\theta_2^1 = \cos^{-1}(< x_2^0, x_3^0 >)$, and thus,

$$x_1^1 = \frac{1}{\Theta_2} x_0^1 + \frac{\Theta_1}{\Theta_2} x_2^0, \tag{3.24}$$

where

$$\Theta_1 = \frac{1}{\left(-\theta_0^1 \cot(\theta_0^1) + \theta_2^1 \cot(\theta_2^1)\right)} \frac{\theta_2^1}{\sin(\theta_2^1)},$$

and

$$\Theta_2 = \frac{1}{\left(-\theta_0^1 \cot(\theta_0^1) + \theta_2^1 \cot(\theta_2^1)\right)} \frac{\theta_0^1}{\sin(\theta_0^1)}.$$

Remark 3.1 Let β_k^i be a sequence of N spherical Bézier curves of order $k \geq 2$, defined by a set of control points $x_k^i, i = 0, ..., N - 1$. In [14], an equivalent necessary and sufficient condition for a spherical Bézier spline $\beta : [O, N] \to \mathbb{S}^n$ defined by

3.3 Spherical Bézier Splines

Fig. 3.2 Illustration of a C^1 Cubic Bézier spline with 3 cubic Bézier curves on $\mathbb{S}^2$

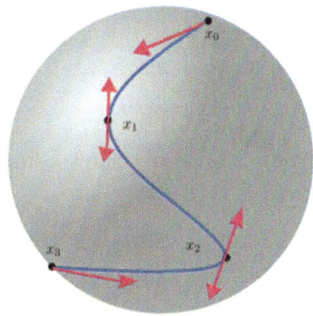

$$\beta(t) = \beta_k^i(t-i; x_0^0, \ldots, x_k^i), \quad \forall t \in [t_i, t_{i+1}] = [i, i+1], \quad i = 0, \ldots, N-1, \tag{3.25}$$

to achieve C^1 continuity is given by the following conditions:

$$x_k^i = x_0^{i+1} \tag{3.26}$$

$$x_1^{i+1} = \varphi_{x_k^i}(x_{N-1}^i) \tag{3.27}$$

where $\varphi_{x_k^i}$ is the geodesic symmetry on $\mathbb{S}^n$ given by Eq. 3.13.

Proposition 3.4 *The spherical Bézier spline $\beta : [0, N] \to \mathbb{S}^n$ defined by (3.18) is C^2 if the following conditions hold*

$$x_{k^i}^i = x_0^{i+1} \tag{3.28}$$

$$k^i \dot{\alpha}(1, x_{k^i-1}^i, x_{k^i}^i) = k^{i+1} \dot{\alpha}(0, x_0^{i+1}, x_1^{i+1}) \tag{3.29}$$

$$k^i(k^i-1)\Omega_i = k^{i+1}(k^{i+1}-1)\Omega_0 \tag{3.30}$$

Here, Ω_i and Ω_0 are defined in Theorem 3.1.

Proof Suppose that β is C^1, the spherical Bézier spline β achieves C^2 continuity if and only if

$$\frac{D}{dt}\Big|_{t=1} \dot{\beta}_k(t; x_0^i, \ldots, x_{k^i}^i) = \frac{D}{dt}\Big|_{t=0} \dot{\beta}_k(t; x_0^{i+1}, \ldots, x_{k^{i+1}}^{i+1}) \quad i = 0, \ldots, N-1. \tag{3.31}$$

Applying Theorem 3.2, we show that (3.31) holds if and only if $k^i(k^i-1)\Omega_i = k^{i+1}(k^{i+1}-1)\Omega_0$. This concludes the proof. □

Proposition 3.5 *Consider a spherical Bézier spline $\beta : [0, N] \to \mathbb{S}^n$ defined by a sequence of N spherical Bézier curves of order $k \geq 2$, denoted as β_k^i, with control points x_k^i for $i = 0, \ldots, N-1$. In [16], achieving C^2 continuity for the spherical Bézier spline is contingent upon satisfying the following equivalent necessary and sufficient conditions:*

$$x_{k^i}^i = x_0^{i+1} \tag{3.32}$$

$$x_1^{i+1} = \varphi_{x_k^i}(x_{k-1}^i) \tag{3.33}$$

$$(d\varphi)_{x_1^{i+1}}(\dot\alpha(0, x_1^{i+1}, x_2^{i+1})) = \dot\alpha(1, x_{k-1}^i, x_{k-1}^i) - 2\dot\alpha(0, x_{k-1}^i, x_k^i) \tag{3.34}$$

Proof The proof follows directly from the application of Lemma 3.3 and Theorem 3.1. □

Algorithm 5 outlines the steps required for constructing a C^2 spherical cubic Bézier spline on $\mathbb{S}^n$.

Algorithm 5 Construction of C^2 spherical cubic Bézier spline
---
Input: $N \in \mathbb{N}$, $z_0, z_1, \ldots, z_{2N-1}$ points on $\mathbb{S}^n$ with $d(z_i, z_{i+1}) < \pi$.
Output: C^2 spherical cubic Bézier spline β.
1: For $i = 1, \ldots, N$ and $j = 0, \ldots, 3N+2$, compute control points x_j^i to define the intermediate Bézier curves β_3^i as follows:
2: Set
$$x_0^1 = z_0, \quad x_1^1 = z_1, \quad x_2^1 = z_2$$
3: Compute the endpoint of each sherical cubic Bézier curve β_3^i
$$x_k^i = z_{i+2}, i = 1, \ldots, N.$$
4: Compute $x_1^{i+1} = \varphi_{x_k^i}(x_{k-1}^i)$, $i = 1, \ldots, N-1$.
5: Compute $x_2^{i+1} = \exp_{x_1^{i+1}}\left((d\varphi)_{x_{k-1}^i}\left(\dot\alpha(1, x_{k-2}^i, x_{k-1}^i) - 2\dot\alpha(0, x_{k-1}^i, x_k^i)\right)\right)$, $i = 1, \ldots, N-1$.
6: **Return** x_j^i, $i = 1, \ldots, N$ and $j = 0, \ldots, 3N+2$ control points of the intermediate N spherical cubic Bézier curves and a C^2 spherical cubic Bézier spline β.

3.4 Solving Interpolation Problem on the Sphere $\mathbb{S}^n$

Motivated by the theoretical insights gained from the two previous sections, this section delves into the interpolation problem (3.1) on $\mathbb{S}^n$, serving as a primary example of a symmetric Riemannian manifold. We consider $x_0, \ldots, x_N$ as a sequence of $(N+1)$ data points in $\mathbb{S}^n$ with x_j being in the cut locus of x_i (i.e. x_i and x_j are two non-antipodal points on $\mathbb{S}^n$). Let $t_0 < \ldots < t_N$ as a sequence of time instants, where, for simplicity, we assume that $t_i = i$. The objective is to construct a C^2 spherical Bézier spline $\gamma : [0, N] \to \mathbb{S}^n$ that interpolates the given set of data points x_i, i.e., $\gamma(t_i) = x_i$ for $i = 0, \ldots, N$, while minimizing the following cost function:

$$\min_{\gamma \in C^2([0,N],\mathbb{S}^n)} E(\gamma) = \int_0^N \left\langle \frac{D^2\gamma(t)}{Dt^2}, \frac{D^2\gamma(t)}{Dt^2} \right\rangle dt. \tag{3.35}$$

3.4 Solving Interpolation Problem on the Sphere $\mathbb{S}^n$

In the construction of the spherical Bézier spline γ, we assume that it is defined by a sequence of N spherical Bézier curves γ_k^i, where $k \in \{2, 3\}$ and $i = 0, \ldots, N-1$. Specifically, we consider the segments joining x_0 and x_1, as well as the segment joining x_{N-1} and x_N, to be spherical Bézier curves of order two. All other segments are spherical Bézier curves of order three. Additionally, let $(\widehat{z_i^-}, \widehat{z_i^+})$ be control points on the left and right-hand side of the interpolation point x_i for $i = 1, \ldots, (N-1)$. The Bézier spline $\gamma : [0, N] \longrightarrow \mathbb{S}^n$ is then given by:

$$\gamma(t) = \begin{cases} \gamma_2^0(t; x_0, \widehat{z_1^-}, x_1), & 0 \leq t \leq 1, \\ \gamma_3^i(t-i; x_i, \widehat{z_i^+}, \widehat{z_{i+1}^-}, x_{i+1}), & i-1 \leq t \leq i, \\ \gamma_2^{N-1}(t-(N-1); x_{N-1}, \widehat{z_{n-1}^+}, x_n), & N-1 \leq t \leq N. \end{cases}$$

The Bézier spline γ ensures interpolation of the first and last control points of each Bézier curve γ_j^i, where $j \in \{2, 3\}$ and $0 \leq i \leq N-1$. Consequently, the continuity of γ at the joint points is adequately satisfied. However, as discussed in the previous section, the ideal scenario is for γ to belong to the class C^2. To achieve this objective, our proposed algorithm in [18] proceeds in two main steps: First, it solves for a C^1 spherical Bézier spline on $\mathbb{S}^n$ and then uses this solution to construct a C^2 spherical Bézier spline.

3.4.1 C^1 Spherical Bézier Spline

Let's begin by outlining the conditions necessary to achieve C^1 continuity along the spherical Bézier spline. The proposed approach involves adapting the interpolation problem to a vector space, similar to the Euclidean case, in order to determine the control points that generate the spline γ. Specifically, the vector space under consideration is the tangent space $T_{x_i}\mathbb{S}^n$, where $i = 1, \ldots, N-1$. Importantly, the presented algorithm relies solely on the Riemannian exponential and logarithm maps.

Given $x_0, \ldots, x_N$ a sequence of $(N+1)$ data points in $\mathbb{S}^n$ with x_j being in the cut locus of x_i, we employ the Riemannian logarithmic map, as defined by Eq. 3.10, to lift data points $x_0, \ldots, x_N$ into each tangent space $T_{x_i}\mathbb{S}^n$, where $i = 1, \ldots, N-1$. The mapped data are then represented by $v^i = (v_0^i, \ldots, v_N^i)$, with $v_k^i = \mathrm{Log}_{x_i}(x_k)$ for $k = 0, \ldots, N$. Let $\beta : [t_0, t_N] \to T_{x_i}\mathbb{S}^n$ denote the Bézier spline on $T_{x_i}\mathbb{S}^n$, $i = 1, \ldots, N-1$, defined by N Bézier curves $\beta_j^i(t, \widehat{w}_0^1, \ldots, \widehat{w}_k^1)$, $j \in \{2, 3\}, 0 \leq i \leq N-1$. Consequently, the optimization problem (3.35) can be reformulated as follows on $T_{x_i}\mathbb{S}^n$:

$$\min_{(\widehat{w}_1^i)^-,\ldots,(\widehat{w}_{N-1}^i)^-} E((\widehat{w}_1^i)^-, \ldots, (\widehat{w}_{N-1}^i)^-) := \min_{(\widehat{w}_1^i)^-,\ldots,(\widehat{w}_{N-1}^i)^-} \int_0^1 \|(\ddot{\beta}_2^i)^0(t; v_0^i, (\widehat{w}_1^i)^-, v_1^i)\|_2^2$$

$$+ \sum_{i=1}^{N-2} \int_0^1 \|\ddot{\beta}_3^i(t; v_i^i, (\widehat{w}_i^i)^+, (\widehat{w}_{i+1}^i)^-, v_{i+1}^i)\|_2^2$$

$$+ \int_0^1 \|\ddot{\beta}_2^{N-1}(t; v_{N-1}^i, (\widehat{w}_{N-1}^i)^+, v_N^i)\|_2^2. \quad (3.36)$$

Solving the optimization problem (3.36) in each tangent space $T_{x_i}\mathbb{S}^n$ yields the control point $(\widehat{z}^i)^-$ of the spherical Bézier spline on $\mathbb{S}^n$. Moreover, by applying Algorithm 3 developed in chapter 1 for the Euclidean case, we establish conditions under which the Bézier curve β is of class C^2. Subsequently, the Riemannian exponential map Exp_{x_i} defined on $\mathbb{S}^n$ by Eq. 3.9 is employed to transition the control points of the C^2 Bézier curve β from the tangent space $T_{x_i}\mathbb{S}^n$ to $\mathbb{S}^n$. Additionally, we demonstrate that the resulting control points on $\mathbb{S}^n$ are optimal, ensuring that the curve is of class C^1.

Theorem 3.2 *Given $x_0, ..., x_N$, a sequence of $(N+1)$ data points on $\mathbb{S}^n$ with x_j lying in the cut locus of x_i for each $i = 1, ..., N-1$. Let $V = [v_0^i, ..., v_N^i]$ denote the corresponding mapped data in the tangent space $T_{x_i}\mathbb{S}^n$ at x_i, defined by $v_k^i = \text{Log}_{x_i}(x_k)$ for $k = 0, ..., N$. Set $0 = t_0 < ... < t_N = 1$ as a sequence of time instants. Then, there exists a unique solution W_i that minimizes Eq. 3.36 and containing the $(N-1)$ control points that generate the C^2 Bézier curve β_i, in each tangent space $T_{x_i}\mathbb{S}^n$, and a matrix $\tilde{X} = [\tilde{x}_0^i, ..., \tilde{x}_N^i]^T$ of size $(n+1)(N+1) \times (n+1)$ containing the new $(N+1)$ interpolation points in each tangent space $T_{x_i}\mathbb{S}^n$.*

Proof We give the proof only for the case $i = 1$, the other ones being similar. Let $x_0, ..., x_N$ be a set of data points on $\mathbb{S}^n$ and $\tilde{V} = [v_0^1, ..., v_N^1]$ their corresponding mapped data in the tangent space $T_{x_1}\mathbb{S}^n$. Let $t \longrightarrow \beta_k(t; \widehat{w}_0^1, ..., \widehat{w}_k^1)$ denote the bézier curve of order $k \in \{2, 3\}$ in $T_{x_1}\mathbb{S}^n$ defined with its control points $\widehat{w}_0^1, ..., \widehat{w}_k^1$ in the Bernstein basis polynomials of degree k as :

$$\beta_2(t; \widehat{w}_0^1, \widehat{w}_1^1, \widehat{w}_2^1) = \widehat{w}_0^1(1-t)^2 + 2\widehat{w}_1^1(1-t)t + \widehat{w}_2^1 t^2, \qquad (3.37)$$
$$\beta_3(t; \widehat{w}_0^1, \widehat{w}_1^1, \widehat{w}_2^1, \widehat{w}_3^1) = \widehat{w}_0^1(1-t)^3 + 3\widehat{w}_1^1 t(1-t)^2 + 3\widehat{w}_2^1 t^2(1-t) + \widehat{w}_3^1 t^3.$$

Let $((\widehat{w}_k^1)^-, (\widehat{w}_k^1)^+)$ denote the two control points on the left and the right hand side of the interpolation point v_k^1, for $k = 1, ..., N-1$. Similarly to the Euclidean case, the differentiability condition (2.36) allows us to express $(\widehat{w}_k^1)^+$ in terms of $(\widehat{w}_k^1)^-$ which leads to an unconstrained optimization problem as a function of $(\widehat{w}_k^1)^-$ only given by :

$$\min_{(\widehat{w}_1^1)^-, ..., (\widehat{w}_{N-1}^1)^-} E((\widehat{w}_1^1)^-, ..., (\widehat{w}_{N-1}^1)^-) := \min_{(\widehat{w}_1^1)^-, ..., (\widehat{w}_{N-1}^1)^-} \int_0^1 \|\ddot{\beta}_2^0(v_0^1, (\widehat{w}_1^1)^-, v_1^1; t)\|^2 dt$$
$$+ \sum_{k=1}^{N-2} \int_0^1 \|\ddot{\beta}_3^k(v_k^1, (\widehat{w}_k^1)^-, (\widehat{w}_{k+1}^1)^-, v_{k+1}^1; t)\|^2 dt$$
$$+ \int_0^1 \|\ddot{\beta}_2^{N-1}(v_{N-1}^1, (\widehat{w}_{N-1}^1)^-, v_N^1; t)\|^2 dt,$$

To solve this optimization problem, we need to compute the critical points of the gradient of the energy function. Let us first examine the inner product of the acceleration. In fact, the acceleration on respective intervals is given by :

3.4 Solving Interpolation Problem on the Sphere $\mathbb{S}^n$

$$\ddot{\beta}_2(t; \widehat{w}_0^1, \widehat{w}_1^1, \widehat{w}_2^1) = 2\widehat{w}_0^1 - 4\widehat{w}_1^1 + 2\widehat{w}_2^1,$$
$$\ddot{\beta}_3(t; \widehat{w}_0^1, \widehat{w}_1^1, \widehat{w}_2^1, \widehat{w}_3^1) = 6\widehat{w}_0^1 - 12\widehat{w}_1^1 + 6\widehat{w}_2^1 + 6t(-\widehat{w}_0^1 + 3\widehat{w}_1^1 - 3\widehat{w}_2^1 + \widehat{w}_3^1).$$
(3.38)

Then we compute the inner product of the acceleration with respect to the inner product defined in $T_{x_1}\mathbb{S}^n$, we thus get:

$$||\ddot{\beta}_2||^2 = 4(<\widehat{w}_0^1, \widehat{w}_0^1> -4<\widehat{w}_0^1, \widehat{w}_1^1> +2<\widehat{w}_0^1, \widehat{w}_2^1> +4<\widehat{w}_1^1, \widehat{w}_1^1>$$
$$-4<\widehat{w}_2^1, \widehat{w}_1^1> + <\widehat{w}_2^1, \widehat{w}_2^1>),$$
$$||\ddot{\beta}_3||^2 = 36 < \widehat{w}_0^1 - 2\widehat{w}_1^1 +_2^1 + t(-\widehat{w}_0^1 + 3\widehat{w}_1^1 - 3\widehat{w}_2^1 + \widehat{w}_3^1), \widehat{w}_0^1 - 2\widehat{w}_1^1 + \widehat{w}_2^1$$
$$+ t(-\widehat{w}_0^1 + 3\widehat{w}_1^1 - 3\widehat{w}_2^1 + \widehat{w}_3^1) >;$$

Finally we evaluate the integral of each term of Eq. 2.40, we obtain:

$$\int_0^1 ||\ddot{\beta}_2||^2 dt = ||\ddot{\beta}_2||^2 = 4(<\widehat{w}_0^1, \widehat{w}_0^1> -4<\widehat{w}_0^1, \widehat{w}_1^1> +2<\widehat{w}_0^1, \widehat{w}_2^1>$$
$$+4<\widehat{w}_1^1, \widehat{w}_1^1> -4<\widehat{w}_2^1, \widehat{w}_1^1> + <\widehat{w}_2^1, \widehat{w}_2^1>),$$

$$\int_0^1 ||\ddot{\beta}_3||^2 dt = 36(\frac{1}{3}<\widehat{w}_0^1, \widehat{w}_0^1> - <\widehat{w}_0^1, \widehat{w}_1^1> +\frac{1}{3}<\widehat{w}_0^1, \widehat{w}_3^1> + <\widehat{w}_1^1, \widehat{w}_1^1>$$
$$- <\widehat{w}_1^1, \widehat{w}_2^1> + <\widehat{w}_2^1, \widehat{w}_2^1> - <\widehat{w}_2^1, \widehat{w}_3^1> +\frac{1}{3}<\widehat{w}_3^1, \widehat{w}_3^1>).$$

By replacing $\widehat{w}_k^1$ such that $\widehat{w}_0^1 = v_0^1$, $\widehat{w}_1^1 = (\widehat{w}_1^1)^-$ and $\widehat{w}_2^1 = v_1^1$ for the first interval, $\widehat{w}_0^1 = v_{n-1}^1$; $\widehat{w}_1^1 = (\widehat{w}_{N-1}^1)^-$ and $\widehat{w}_2^1 = v_N^1$ for the last interval; and $\widehat{w}_0^1 = v_k^1$, $\widehat{w}_1^1 = (\widehat{w}_k^1)^+$, $\widehat{w}_2^1 = (\widehat{w}_{k+1}^1)^-$, and $\widehat{w}_3^1 = v_{k+1}^1$ for the intermediate intervals, it leads to the apparition of constant terms K involving all v_k^1 which will disappear when we derive the gradient.

As in the proof of Theorem 2.2, the new formulation becomes:

$$\min_{(\widehat{w}_1^1)^-, \dots, (\widehat{w}_{N-1}^1)^-} E((\widehat{w}_1^1)^-, \dots, (\widehat{w}_{N-1}^1)^-) = \tag{3.39}$$

$$4(-4<v_0^1, (\widehat{w}_1^1)^-> +4<(\widehat{w}_1^1)^-, (\widehat{w}_1^1)^-> -4<v_0^1, (\widehat{w}_1^1)^->) \tag{3.40}$$

$$+ \sum_{k=1}^{N-2} 36(- <v_k^1, (\widehat{w}_k^1)^+> + <(\widehat{w}_k^1)^+, \widehat{w}_k^1)^+> - <(\widehat{w}_k^1)^+, (\widehat{w}_{k+1}^1)^->$$
(3.41)

$$+ \leq (\widehat{w}_{k+1}^1)^-, (\widehat{w}_{k+1}^1)^-> - <(\widehat{w}_{k+1}^1)^-, v_{k+1}^1>) \tag{3.42}$$
$$+ 4(-4<v_{N-1}^1, (\widehat{w}_{N-1}^1)^+> +4<(\widehat{w}_{N-1}^1)^+, (\widehat{w}_{N-1}^1)^+> -4<v_N^1, (\widehat{w}_{N-1}^1)^+>)$$
$$+ K \tag{3.43}$$

After changing $(\widehat{w}_k^1)^+$'s by $(\widehat{w}_k^1)^-$'s using Eqs. 2.37–2.39 the next step is to compute the gradient and search for the optimal solution $(\widehat{w}_k^1)^-$. Consequently, $\forall u \in T_{x_1} \mathbb{S}^n$ the optimal solution is given by:

$$\frac{\partial E}{\partial (\widehat{w}_1^1)^-} = 4(8 < (\widehat{w}_1^1)^-, u > -4 < v_0^1, u > -4 < v_1^1, u >)$$
$$+ 36(-\frac{14}{9} < v_1^1, u > +\frac{8}{9} < \widehat{w}_1^1)^-, u > +\frac{6}{9} < \widehat{w}_2^1)^-, u >) = 0,$$

$$\frac{\partial E}{\partial (\widehat{w}_2^1)^-} = 36(-\frac{5}{3} < v_1^1, u > +\frac{2}{3} < (\widehat{w}_1^1)^-, u > +2 < (\widehat{w}_2^1)^-, u > - < v_2^1, u >)$$
$$+ 36(-3 < v_2^1, u > +2 < (\widehat{w}_2^1)^-, u > + < (\widehat{w}_3^1)^-, u >) = 0,$$

$$\frac{\partial E}{\partial (\widehat{w}_j^1)^-} = 36(-3 < v_j^1, u > +2 < (\widehat{w}_j^1)^-, u > + < (\widehat{w}_{j+1}^1)^-, u >)$$
$$+ 36(-2 < v_{j-1}^1, u > + < (\widehat{w}_{j-1}^1)^-, u > +2 < (\widehat{w}_{j-1}^1)^-, u > - < v_j^1, u >) = 0 \; j = 2, ..., N-2,$$

$$\frac{\partial E}{\partial (\widehat{w}_{N-1}^1)^-} = 36(-2 < v_{N-2}^1, u > + < (\widehat{w}_{N-2}^1)^-, u > +2 < (\widehat{w}_{N-1}^1)^-, u > - < v_{N-1}^1, u >)$$
$$+ 4(-24 < v_{N-1}^1, u > +18 < (\widehat{w}_{N-1}^1)^-, u > +6 < \widehat{w}_N^1, u >) = 0.$$

It follows that, similar as in the proof of Theorem 2.2, the solution of Eq. 2.42 is given by

$$\widehat{W}_1 = [(\widehat{w}_1^1)^-, ..., (\widehat{w}_{N-1}^1)^-]^T = DV$$

Analogous to the Euclidean space, with this preliminary step completed, we can deduce the matrix $W_1 = [(w_1^1)^-, ..., (w_{N-1}^1)^-]^T$ containing the control points of the C^2 Bézier curve β_1 in $T_{x_1}\mathbb{S}^n$. Therefore, we obtain:

$$(w_1^1)^- = (\widehat{w}_1^1)^-, \tag{3.44}$$

$$(w_2^1)^- = \frac{1}{3}v_0^1 - \frac{1}{2}(w_1^1)^- + \frac{8}{3}v_1^1, \tag{3.45}$$

$$(w_{i+1}^1)^- = (w_{i-1}^1)^+ + 4v_i^1 - 4(w_i^1)^-, i = 2, ..., N-2. \tag{3.46}$$

and the new $(N + 1)$ interpolation points $\tilde{V} = (\tilde{v}_0^1, ..., \tilde{v}_N^1)$ on $T_{x_1}\mathbb{S}^n$ are given by:

$$\tilde{v}_k^1 = v_k^1, \text{ for } k = 0, .., N-1, \tag{3.47}$$

$$\tilde{v}_N^1 = 2v_{N-1}^1 + 2(w_{N-1}^1)^+ - 6(w_{N-1}^1)^- + 3(w_{N-2}^1)^+, \tag{3.48}$$

which completes the proof. $\square$

Proposition 3.6 *Under the same hypotheses detailed in Theorem 3.2, there exists a unique matrix* $\widehat{Z} = [\widehat{z_1}, ..., \widehat{z_{N-1}}]^T \in \mathbb{R}^{(n+1)(N-1) \times (n+1)}$, *containing the* $(N-1)$ *control points that generate the spherical Bézier spline* γ *interpolating the points* x_i *at* t_i *on* $\mathbb{S}^n$, *for* $i = 0, ..., N$. *The rows of* $\widehat{Z}$ *are given by:*

3.4 Solving Interpolation Problem on the Sphere $\mathbb{S}^n$

$$\widehat{z_i^-} = exp_{x_i}(w_i), \ i = 1, ..., N-1, \tag{3.49}$$

where w_i, represent the row i of W_i in $T_{x_i}\mathbb{S}^n$, for $i = 1, .., N-1$. Moreover, the new $(N+1)$ interpolation points in $\mathbb{S}^n$ are given by:

$$\tilde{x}_k = exp_{x_i}(\tilde{v}_k^i), \ k = 0, ..., N; \ i = 1, ..., N-1. \tag{3.50}$$

Corollary 3.1 *The spherical Bézier spline $\gamma : [0, 1] \to \mathbb{S}^n$ is C^1 on $\mathbb{S}^n$.*

Proof For $i = 0, ..., N-2$, by construction, we have $\gamma|_{[t_i,t_{i+1}]}$ is smooth. Furthermore, in a Euclidean space, the Bézier curve β is C^1 under the differentiability condition 2.36. This assumption allows us to express $\widehat{z_i^+}$ in terms of $\widehat{z_i^-}$ in the form

$$\widehat{z_i^+} = x_i + \rho_i(\widehat{z_i^-} - x_i) \tag{3.51}$$

Moreover, in a Euclidean space we have: $\log_x(\hat{z}) = \hat{z} - x$. This allows us to generalize the Eq. 3.51 to $\mathbb{S}^n$ in the form $\widehat{z_i^+} = exp_{x_i}(\rho_i \exp_{x_i}^{-1}(\widehat{z_i^-}))$, for $i = 1, ..., N-1$ and ensure that γ is C^1 on $\mathbb{S}^n$. □

The algorithm for constructing a C^1 Bézier spline on $\mathbb{S}^n$ is outlined in Algorithm 6, including all the necessary steps.

Algorithm 6 Construction of C^1 interpolating spherical Bézier spline on $\mathbb{S}^n$.

Input: $N \geq 3$, $X = [x_0, ..., x_N]^T$ a matrix of size $(n+1)(N+1) \times (n+1)$ containing the $(N+1)$ interpolation points on $\mathbb{S}^n$.
Output: $\widehat{Z}$ and $\tilde{X}$.
1: **for** $i = 1 : N-1$ **do**
2:   Calculate $V = [v_0^i, ..., v_N^i]^T$ a matrix of size $(n+1)(N+1) \times (n+1)$ containing the $(N+1)$ interpolation points on $T_{x_i}\mathbb{S}^n$:
3:   **for** $k = 0 : N$ **do**
4:     $v_k^i = \text{Log}_{x_i}(x_k)$
5:   Calculate $W_i = [(w_1^i)^-, ..., (w_{N-1}^i)^-]^T$ a matrix of size $(n+1)(N-1) \times (n+1)$ containing the $(N-1)$ control points of the C^2 Bézier curve β_i on $T_{x_i}\mathbb{S}^n$, and $\tilde{V} = [\tilde{v}_0^i, ..., \tilde{v}_N^i]^T$ a matrix of size $(n+1)(N+1) \times (n+1)$ containing the new interpolation points on $T_{x_i}\mathbb{S}^n$ using Algorithm 4.
6:     Calculate control point $\widehat{z_i^-}$ with $\widehat{z_i^-} = \text{Exp}_{x_i}((w_i^i)^-)$
7:     Calculate the new interpolation points $\tilde{x}_k = \text{Exp}_{x_i}(\tilde{x}_k^i)$.
8:   **end for**
9: **end for**
10: **return** $\widehat{Z}$ and $\tilde{X}$.

3.4.2 C^2 Spherical Bézier Spline

Assuming γ is C^1, and having obtained the solution $\widehat{Z}$, let's denote the new control points on the left and right sides of the interpolation point $\tilde{x}_i$ that generate the C^2

spherical Bézier spline γ on $\mathbb{S}^n$ as z_i^- and z_i^+, respectively. The process of finding the control points z_i^- for $i = 1, ..., N - 1$ follows a similar approach to the Euclidean case. Specifically, we may already know z_1^- (and consequently z_1^+ due to the C^1 differentiability condition on $\mathbb{S}^n$) and aim to iteratively define z_i^- for $i = 2, ..., N - 1$ (as well as z_i^+ similar to z_1^+).

Theorem 3.3 *Let $\tilde{x}_0, ..., \tilde{x}_N$ constitute a set of distinct points $\mathbb{S}^n$ as given by (3.50), and let $\alpha(t)$ be the shortest geodesic arc joining the control points of the curve γ on $\mathbb{S}^n$ defined by Eq. 3.7. Consider $W_1 = [(w_1^1)^-, ..., (w_{N-1}^1)^-]^T$, the matrix of size $(n+1)(N-1) \times (n+1)$ containing the control points of the C^2 Bézier curve β_1 in $T_{x_1}\mathbb{S}^n$. Then, there exists a unique matrix $Z = [z_1^-, ..., z_{N-1}^-]^T \in \mathbb{R}^{(n+1)(N-1) \times (n+1)}$, containing the $(N - 1)$ control points that generate the C^2 spherical Bézier spline γ, interpolating the points $\tilde{x}_i$ at t_i on $\mathbb{S}^n$, for $i = 0, ..., N$. The rows of Z are given by:*

1. $z_1^- = Exp_{\tilde{x}_1}((w_1^1)^-)$,
2. $z_2^- = Exp_{z_1^+}\left(\frac{1}{3}\left((d\varphi_{\tilde{x}_1})_{z_1^-}\left(\dot{\alpha}(1, \tilde{x}_0, z_1^-)\right) - 4\dot{\alpha}(0, z_1^-, \tilde{x}_1)\right)\right)$,
3. $z_{i+1}^- = Exp_{z_i^+}\left(\left((d\varphi_{\tilde{x}_i})_{z_i^-}\left(\dot{\alpha}(1, z_{i-1}^+, z_i^-)\right) - 2\dot{\alpha}(0, z_i^-, \tilde{x}_i)\right)\right)$, $i = 2, ..., N - 2$.

We will use an adapted version of Theorem 3.1 and Theorem 2 in [16] to derive the proof of Theorem 3.3.

Proof Part (1) is a consequence of Proposition 7.1. Now, we proceed to prove (2). The Bézier curve γ is C^2 on $\mathbb{S}^n$ if and only if it satisfies the C^2 differentiability condition at joint points $\tilde{x}_i$ for $i = 1, ..., N - 1$. Specifically, at the point $\tilde{x}_1$, this condition can be expressed as follows:

$$\frac{D}{dt}\Big|_{t=1} \dot{\gamma}_2(t; \tilde{x}_0, z_1^-, \tilde{x}_1) = \frac{D}{dt}\Big|_{t=0} \dot{\gamma}_3(t; \tilde{x}_1, z_1^+, z_2^-, \tilde{x}_2). \quad (3.52)$$

Using Theorem 3.1, we can establish that γ is C^2 at $\tilde{x}_1$ if and only if $\Omega_2 - 3\Omega_0 = 0$, where:

$$\Omega_2 - 3\Omega_0 = (d\text{Exp}_{\tilde{x}_1})^{-1}_{-\dot{\alpha}(1,z_1^-,\tilde{x}_1)} \left(\dot{\alpha}(0, z_1^-, \tilde{x}_1) - \dot{\alpha}(1, \tilde{x}_0, z_1^-)\right)$$
$$- 3(d\text{Exp}_{\tilde{x}_1})^{-1}_{\dot{\alpha}(0,\tilde{x}_1,z_1^+)} \left(\dot{\alpha}(0, z_1^+, z_2^-) - \dot{\alpha}(1, \tilde{x}_1, z_1^+)\right). \quad (3.53)$$

As β_1 is a C^1 spherical Bézier spline on $T_{x_1}\mathbb{S}^n$, we can assert that: $\dot{\alpha}(1, z_1^-, \tilde{x}_1) = \dot{\alpha}(0, \tilde{x}_1, z_1^+)$. Referring to Lemma 3.3, we obtain:

$$(d\text{Exp}_{\tilde{x}_1})^{-1}_{\dot{\alpha}(0,\tilde{x}_1,z_1^+)} \left(\dot{\alpha}(0, z_1^+, z_2^-) - \dot{\alpha}(1, \tilde{x}_1, z_1^+)\right)$$
$$= -(d\text{Exp}_{\tilde{x}_1})^{-1}_{-\dot{\alpha}(0,\tilde{x}_1,z_1^+)} \left((d\varphi_{\tilde{x}_1})_{z_1^+}\left(\dot{\alpha}(0, z_1^+, z_2^-) - \dot{\alpha}(1, \tilde{x}_1, z_1^+)\right)\right).$$

3.4 Solving Interpolation Problem on the Sphere $\mathbb{S}^n$

It follows that:

$\Omega_2 - 3\Omega_0$

$= (d\mathrm{Exp}_{\tilde{x}_1})^{-1}_{-\dot\alpha(0,\tilde{x}_1,z_1^+)} \left(\dot\alpha(0, z_1^-, \tilde{x}_1) - \dot\alpha(1, \tilde{x}_0, z_1^-) \right)$

$+ 3(d\mathrm{Exp}_{\tilde{x}_1})^{-1}_{-\dot\alpha(0,\tilde{x}_1,z_1^+)} \left((d\varphi_{\tilde{x}_1})_{z_1^+} \left(\dot\alpha(0, z_1^+, z_2^-) - \dot\alpha(1, \tilde{x}_1, z_1^+) \right) \right)$

$= (d\mathrm{Exp}_{\tilde{x}_1})^{-1}_{-\dot\alpha(0,\tilde{x}_1,z_1^+)} \Bigg[3(d\varphi_{\tilde{x}_1})_{z_1^+} \left(\dot\alpha(0, z_1^+, z_2^-) \right) - 3(d\varphi_{\tilde{x}_1})_{z_1^+} \left(\dot\alpha(1, \tilde{x}_1, z_1^+) \right)$

$+ \dot\alpha(0, z_1^-, \tilde{x}_1) - \dot\alpha(1, \tilde{x}_0, z_1^-) \Bigg]$

Hence, $\Omega_2 - 3\Omega_0 = 0$, if and only if

$$3(d\varphi_{\tilde{x}_1})_{z_1^+} \left(\dot\alpha(0, z_1^+, z_2^-) \right) - 3(d\varphi_{\tilde{x}_1})_{z_1^+} \left(\dot\alpha(1, \tilde{x}_1, z_1^+) \right) + \dot\alpha(0, z_1^-, \tilde{x}_1) - \dot\alpha(1, \tilde{x}_0, z_1^-) = 0. \qquad (3.54)$$

However, $\varphi_{\tilde{x}_1}\left(\alpha(t, \tilde{x}_1, z_1^+)\right) = \alpha(1-t, z_1^-, \tilde{x}_1), ; \forall t \in [0, 1]$. Upon differentiating this identity with respect to t, we get:

$$(d\varphi_{\tilde{x}_1})_{z_1^+} \left(\dot\alpha(1, \tilde{x}_1, z_1^+) \right) = -\dot\alpha(0, z_1^-, \tilde{x}_1).$$

Accordingly, (3.54) becomes

$$3(d\varphi_{\tilde{x}_1})_{z_1^+} \left(\dot\alpha(0, z_1^+, z_2^-) \right) = \dot\alpha(1, \tilde{x}_0, z_1^-) - 4\dot\alpha(0, z_1^-, \tilde{x}_1). \qquad (3.55)$$

Now, Lemma 3.3 demonstrates that

$$(d\varphi_{\tilde{x}_1})_{z_1^+} \left(\dot\alpha(0, z_1^+, z_2^-) \right) = (d\varphi_{\tilde{x}_1})_{\varphi_{\tilde{x}_1}(z_1^-)} \left(\dot\alpha(0, z_1^+, z_2^-) \right)$$
$$= (d\varphi_{\tilde{x}_1})^{-1}_{z_1^-} \left(\dot\alpha(0, z_1^+, z_2^-) \right).$$

$(d\varphi_{\tilde{x}_1})^{-1}_{z_1^-} \left(\dot\alpha(0, z_1^+, z_2^-) \right) = \frac{1}{3} \left(\dot\alpha(1, \tilde{x}_0, z_1^-) - 4\dot\alpha(0, z_1^-, \tilde{x}_1) \right)$. Consequently, using the exponential map at the point z_1^+, we get

$$z_2^- = \mathrm{Exp}_{z_1^+} \left(\frac{1}{3} \left((d\varphi_{\tilde{x}_1})_{z_1^-} \left(\dot\alpha(1, \tilde{x}_0, z_1^-) \right) - 4\dot\alpha(0, z_1^-, \tilde{x}_1) \right) \right). \qquad (3.56)$$

The proof of Part (3) follows in much the same way as Part (2). $\qquad\square$

Remark 3.2 Similar to the Euclidean case, the C^2 differentiability condition at the point $\tilde{x}_{N-1} \in \mathbb{S}^n$, given by

$$\frac{D}{dt}|_{t=1} \dot{\gamma}_3(t; \tilde{x}_{N-2}, z^+_{N-2}, z^-_{N-1}, \tilde{x}_{N-1}) = \frac{D}{dt}|_{t=0} \dot{\gamma}_2(t; \tilde{x}_{N-1}, z^+_{N-1}, \tilde{x}_N), \quad (3.57)$$

will modify the interpolation point $\tilde{x}_N$.

Algorithm 7 Construction of C^2 interpolating spherical Bézier spline on $\mathbb{S}^n$.

Input: $N \geq 3$, $\tilde{X} = [\tilde{x}_0, ..., \tilde{x}_N]^T$ a matrix of size $(n+1)(N+1) \times (n+1)$ containing the $(N+1)$ interpolation points on $\mathbb{S}^n$.
Output: Z.
1: Calculate $\widehat{Z} = [\widetilde{z_1^-}, ..., \widetilde{z_{N-1}^-}]^T$ using Algorithm 6.
2: Set $z_1^- = \widetilde{z_1^-}$.
3: Calculate control point z_1^+:
4: $z_1^+ = \operatorname{Exp}_{\tilde{x}_1}(-\frac{2}{3}\operatorname{Exp}_{\tilde{x}_1}^{-1}(z_1^-))$
5: Calculate control point z_2^-:
6: $z_2^- = \operatorname{Exp}_{z_1^+}\left(\frac{1}{3}\left((d\varphi_{\tilde{x}_1})_{z_1^-}\left(\dot{\alpha}(1, \tilde{x}_0, z_1^-)\right) - 4\dot{\alpha}(0, z_1^-, \tilde{x}_1)\right)\right)$
7: **for** $i = 2 : N-2$ **do** do
8: $\quad z_i^+ = \operatorname{Exp}_{\tilde{x}_i}(-\operatorname{Exp}_{\tilde{x}_i}^{-1}(z_i^-))$
9: $\quad z_{i+1}^- = \operatorname{Exp}_{z_i^+}\left(\left((d\varphi_{\tilde{R}_i})_{z_i^-}\left(\dot{\alpha}(1, z_{i-1}^+, z_i^-)\right) - 2\dot{\alpha}(0, z_i^-, \tilde{x}_i)\right)\right)$
10: **end for**
11: Calculate control point z_{N-1}^+:
12: $z_{N-1}^+ = \operatorname{Exp}_{\tilde{x}_{N-1}}(-\frac{2}{3}\operatorname{Exp}_{\tilde{x}_{N-1}}^{-1}(z_{N-1}^-))$
13: **return** Z,

References

1. Svantesson, M. and Olausson, H. and Eklund, A. and Thordstein, M.: Virtual EEG-electrodes: Convolutional neural networks as a method for upsampling or restoring channels, Journal of Neuroscience Methods, 355, 2021.
2. Leweke, S. and Hauk, O. and Michel, V.: Vector-valued spline method for the spherical multiple-shell electro-magnetoencephalography problem, Inverse Problems, 38 (8), 2022.
3. Qian, W. and Le,P. and Jie-lin, Z.: The construction of spherical interpolation splines with local properties, Journal of Graphics, 42(2): 230-236, 2021.
4. Mardia, K.V. and Jupp, P.: Directional Statistics, 2nd edition John Wiley and Sons, 2000.
5. Shoemake, K.: Quaternion calculus and fast animation, SIGGRAPH'87 Course Notes on State of the Art Image Synthesis, ACM, New York, 101-121, 1987.
6. Machado, L., Silva Leite, F. and Huper, K.: Riemannian means as solutions of variational problems, LMS J. Comput. Math., 9:86-103 (electronic), 2006.
7. Machado, L. and Silva Leite, F.: Fitting smooth paths on Riemannian manifolds, Int. J. Appl. Math. Stat., 4(J06):255-3, 2006.
8. Noakes, L., Heinzinger,G. and Paden, B.: Cubic splines on curved spaces. IMA J. Math. Control Inform., 6(4):465-473, 1989.

References

9. Machado, L. and Monteiro, M.T.T.: Geodesic regression on spheres from a numerical optimization viewpoint, International Journal of Computer Mathematics, 92 (9), 1935-1945, 2015.
10. Machado, L. and Monteiro, M.T.T.: A numerical optimization approach to generate smoothing spherical splines, Journal of Geometry and Physics, 111, 71-81, 2017.
11. Jupp, P. E. and Kent, J. T.: Fitting smooth path to spherical data, Journal of Applied Statistics, 36, 34-46, 1987.
12. Kume, A. and Dryden, I.L and Le, H.: Shape space smoothing splines for planar landmark data, Biometrika, 94, 513-528, 2007.
13. Crouch, P., Kun, G. and Silva Leite, F.: The de Casteljau algorithm on Lie groups and spheres, J. Dyn. Control Syst. 5, 397-429, 1999.
14. Popiel, T. and Noakes, L.: C^2 spherical Bézier splines, Comput. Aided Geom. Des. 23, 261-275, 2006.
15. Bak, K.Y and Shin, J.K and Koo, J.Y: Intrinsic spherical smoothing method based on generalized Bézier curves and sparsity inducing penalization, Journal of Applied Statistics, 50, 9, 1942-1961, 2023.
16. Popiel, T. and Noakes, L.: Bézier curves and C^2 interpolation in Riemannian manifolds, J. Approx. Theory, 148(2), 111-127, 2007.
17. Hüper,K. and Leite, F.S.: On the geometry of rolling and interpolation curves on S^n, $SO(n)$ and Grassmann manifolds, J. Dyn. Control Syst, 13, 467â€"502, 2007.
18. Adouani, I. and Samir, C.: A constructive approximation of interpolating Bézier curves on Riemannian symmetric spaces. J. Optim. Theory. Appl. 187(1), 158-180, 2020.

Chapter 4
Spline Interpolation on the Special Orthogonal Group $SO(n)$

In the preceding chapter, we developed and scrutinized a methodology for curve fitting to data on the sphere. The spherical manifold proved advantageous due to its inherent visualizability, enabling an intuitive understanding of the proposed algorithm. While $\mathbb{S}^2$ served as an excellent preliminary example to assess the concepts, the need to extend the work to more intricate manifolds becomes imperative to substantiate the complexity of the approach.

In this chapter, the focus shifts to investigating curve fitting on a manifold that is challenging to visualize yet practically significant: the compact Lie group $SO(n)$ equipped with a Riemannian metric. This scenario arises in applications involving resampling or denoising data points comprising rotation matrices measured at different times [1–3]. These rotation matrices typically represent the spatial alignment of tangible entities, such as satellites, spacecraft, or robotic arms. Consequently, there has been a persistent demand for geometrically meaningful and computationally efficient smooth splines within this manifold. We extend the methodology presented in Chap. 2 to construct a C^2 interpolating Bézier spline on $SO(n)$ by generalizing the De Casteljau algorithm to $SO(n)$ and leveraging the property that $SO(n)$ is a symmetric Riemannian manifold. The effectiveness of the method is validated through experiments on $SO(3)$, affirming Bézier spline regression as a robust and versatile tool for $SO(n)$-valued regression.

4.1 Problem Formulation

Consider $A_0, \ldots, A_N$ as a set of $(N + 1)$ distinct points on $SO(n)$ corresponding to measurement times $0 = t_0 < t_1 < \cdots < t_N = N$. The objective is to find a curve $\gamma : [0, N] \to SO(n)$ that interpolates the data points A_i, $i = 0, \ldots, N$ and minimizes the following function:

$$E(\gamma) = \frac{1}{2} \int_{t_0}^{t_N} < \frac{D^2\gamma(t)}{Dt^2}, \frac{D^2\gamma(t)}{Dt^2} > dt$$

where $<,>$ corresponds to the Riemannian metric on $SO(n)$. The challenge of constructing smooth interpolating curves in $SO(n)$ is crucial in various applications, particularly in handling rotations and orientation data. For instance, when dealing with the rotation group $SO(3)$, smooth interpolation has direct applications in computer graphics and the animation of 3D objects [1, 3, 4]. Moreover, it plays a vital role in diverse applications such as robot motion planning and machine vision [5, 6]. The relevance of such applications motivates us to explore efficient methods for generating smooth interpolating curves on $SO(n)$.

The construction of interpolating curves on the special orthogonal group $SO(n)$ has been the focus of various studies. One of the pioneering approaches is attributed to Shoemake [7], particularly in the context of $SO(3)$, where he employs a reparametrization of rotation matrices based on the unit quaternion representation. Shoemake's method can be seen as an extension of de Casteljau's algorithm for Bézier curves to $SU(2)$, interpolating two elements of $SO(3)$ through the geodesic that connects them. However, despite its computational efficiency, this algorithm is susceptible to variations based on the chosen local system coordinates. Subsequently, several researchers, including Barr et al. [8], Hart et al. [9], Ge and Ravani [10], and Nielson et al. [11], conducted a more intricate geometric analysis of the unit-quaternion-based method, taking into account the Shoemake algorithm. Despite their capability to produce intrinsic curves, these approaches lack generalization to higher-dimensional manifolds. In a distinct approach, Rodrigues proposed a modified version of the de Casteljau algorithm in [12], constructing C^k-polynomial splines in three steps. However, the resulting solutions involve intricate conditions and are contingent on the reparameterization of $[t_0, t_N]$. Recent advancements in interpolation algorithms for rotations within $SO(n)$ utilize approaches based on the geodesic distance in $SO(n)$ and the high-dimensional Kuramoto model. For in-depth exploration and detailed methodologies, interested readers can refer to the following Refs. [13–15].

In this chapter, we expand upon the method introduced in Chap. 2 to construct a C^2 interpolating Bézier spline on $SO(n)$ equipped with the Frobenius inner product. In [16], a C^2 continuity condition was developed for symmetric Riemannian manifolds, involving the computation of velocities and covariant derivatives at each joint point. However, the computation of the covariant derivative becomes intricate due to the involvement of the inverse of the derivative of the Riemannian exponential map. Building on this work, Geir et al. in [17] simplified the C^2 continuity condition presented in [16] for specific cases of Riemannian symmetric spaces.

This chapter further extends this toolbox by providing explicit solutions that ensure the C^2 differentiability condition at joint points. Leveraging global symmetries, we derive equations for control points. Building on the prior work in Chap. 1, we use these equations to find a C^1 interpolating Bézier curve on the Riemannian manifold. Subsequently, these results are employed to derive explicit formulas for the control points of a C^2 interpolating Bézier curve on $SO(n)$. The proposed method is demonstrated to possess several favorable properties, and the solution is unique in many common situations. The performance of the approach is showcased with potential applications in $SO(3)$.

4.2 Geometry of $SO(n)$

The aim of this section is to revisit certain definitions and properties related to the special orthogonal group $SO(n)$ that will be subsequently used. These fundamental notions are extensively covered in [18, 19].

Let $M_n(\mathbb{R})$ denote the Euclidean space of all real $n \times n$ matrices. We will denote elements of $SO(n)$ as real $n \times n$ orthogonal matrices with unit determinant:

$$SO(n) = \{A \in M_n(\mathbb{R}) \mid A^T A = I, \det(A) = 1\}. \tag{4.1}$$

Lemma 4.1 *The tangent space to $SO(n)$ at the identity $I_n \in SO(n)$ is the set of $n \times n$ skew-symmetric matrices $so(n)$ defined by*

$$T_I SO(n) = \{H \in M_n(\mathbb{R}) : H^T + H = 0\}. \tag{4.2}$$

Proof Consider a smooth curve $\gamma : (-1, 1) \to SO(n)$ with $\gamma(0) = I$. Since it is a curve in $SO(n)$, for each $s \in (-1, 1)$, the following holds: $\gamma(s)\gamma(s)^t = I$. Differentiating the latter equation, we obtain: $\gamma'(s)\gamma(s)^t + \gamma(s)\gamma'(s)^t = 0$. At $s = 0$, this simplifies to:

$$\gamma'(0)\gamma(0)^t + \gamma(0)\gamma'(0)^t = 0$$
$$\gamma'(0)I + I\gamma'(0)^t = 0$$
$$\gamma'(0) + \gamma'(0)^t = 0$$

This implies that $\gamma'(0)$ is a skew-symmetric matrix. Therefore, every tangent vector to $SO(n)$ at I is a skew-symmetric matrix. Thus, $T_I SO(n) \subset so(n)$. Considering that both are vector spaces with dimensions $n(n-1)/2$, it follows that $T_I SO(n) = so(n)$. □

Consequently, at point $A \in SO(n)$, the tangent space to $SO(n)$ at A can be deduced easily as $T_A SO(n) = (L_A)_*(T_I SO(n))$, where $(L_A)_*$ represents the derivative of the map $L_A : SO(n) \to SO(n)$, $B \to A.B$. Therefore,

$$T_A SO(n) = A \cdot so(n) = \left\{ H \in M_n(\mathbb{R}) : AH^T + HA^T = 0 \right\}. \tag{4.3}$$

We equip $SO(n)$ with a bi-invariant metric derived from the Frobenius inner product on $\mathbb{R}^{n \times n}$. In this context, consider $A \in SO(n)$ and $H_1, H_2 \in T_A SO(n)$. The inner product on $T_A SO(n)$ is expressed as:

$$\langle H_1, H_2 \rangle_A = \text{trace}(H_1^T H_2). \tag{4.4}$$

In terms of this bi-invariant metric, the geodesics on $SO(n)$ are determined by the translations of one-parameter subgroups. Specifically, the geodesic curve $t \longrightarrow \alpha(t)$ starting at $A \in SO(n)$ in the direction of $H \in T_{I_n}SO(n)$ takes the form $\alpha(t) = A\exp(tH)$, where $t \in \mathbb{R}$. At the identity matrix I, the Riemannian exponential map $\text{Exp}_I : T_I SO(n) \longrightarrow SO(n)$ and Riemannian logarithm map $\text{Log}_I(A) : SO(n) \to T_I SO(n)$ correspond to the standard exponential and logarithm defined by:

$$\text{Exp}_I(H) = \exp(H) = \sum_{k=0}^{\infty} \frac{H^k}{k!}, \tag{4.5}$$

$$\text{Log}_I(A) = \log(A) = \sum_{k=1}^{\infty} \frac{(-1)^{k+1}}{k}(A - I_n)^k. \tag{4.6}$$

The map log converges for $\|A - I_n\| < 1$. However, the local and multivalued nature of the function log poses significant challenges for interpolation. Nevertheless, if a nonsingular matrix $A \in \mathbb{R}^{n \times n}$ does not have eigenvalues on $\mathbb{R}_0^-$, then there exists a unique real logarithm of A whose eigenvalues lie in $\{z \in \mathbb{C}: \text{Im}(z)| < \pi\}$. As multiplying by an orthogonal matrix preserves isometry, to compute the Riemannian exponential map at a point $A \in SO(n)$, we can transport the tangent vector H from $T_A SO(n)$ to $T_I SO(n)$, compute the exponential at the identity using the matrix exponential, and then transport the result back. In other words, the Riemannian exponential $\text{Exp}_A : T_A SO(n) \longrightarrow SO(n)$ is defined by:

$$\text{Exp}_A(H) = A\text{Exp}_I(A^T H) = A\exp(A^T H), \quad H \in T_A SO(n), \tag{4.7}$$

and its inverse is given by:

$$\text{Log}_{A_1}(A_2) = A_1 \log(A_1^T A_2), \quad A_1, A_2 \in SO(n). \tag{4.8}$$

Therefore, the explicit parameterization for the shortest geodesic arc connecting A_1 and A_2 in $SO(n)$ is as follows:

$$\alpha(t, A_1, A_2) = A_1 \exp(t \log(A_1^T A_2)), \quad t \in [0, 1]. \tag{4.9}$$

Example 4.1 (*Geodesic curve on $SO(3)$*) Let S be a skew-symmetric matrix defined by a vector $s = (s_0, s_1, s_2) \in \mathbb{R}^3$ as follows:

$$S = \begin{pmatrix} 0 & -s_2 & s_1 \\ s_2 & 0 & -s_0 \\ s_1 & s_0 & 0 \end{pmatrix}$$

If s is not the zero vector, then the angle of rotation is $\theta = \|s\|_{\mathbb{R}^3}$, and the corresponding rotation matrix $A \in SO(3)$ is given by

4.2 Geometry of SO(n)

$$A = \exp(S) = \sum_{k=0}^{\infty} \frac{S^k}{k!} = I_3 + \left(\frac{\sin(\theta)}{\theta}\right) S + \left(\frac{1-\cos(\theta)}{\theta^2}\right) S^2. \quad (4.10)$$

The infinite sum reduces to a finite sum because of the identity $S^3 = -\theta^2 S$. For practical purposes, constrain $\theta \in]0, \pi[$. Taking into account that $S^T = -S$, we can compute $\log(A) = S$ as follows:

$$S = \log(A) = \frac{\theta}{2\sin(\theta)}(A - A^T). \quad (4.11)$$

Now, given two rotation matrices A_1 and A_2, by Eq. 4.9, the shortest geodesic connecting them is

$$\alpha(t, A_1, A_2) = A_1 \exp(t \log(A_1^T A_2)), \quad t \in [0, 1].$$

If we denote $A = A_1^T A_2$ and use Eqs. 4.10 and 4.11 computed previously, we obtain that

$$\alpha(t, A_1, A_2) = \left[I_3 + \left(\frac{\sin(t\theta)}{\theta}\right) S + \left(\frac{1-\cos(t\theta)}{\theta^2}\right) S^2 \right] A_1. \quad (4.12)$$

It is easy to check that $\alpha(0, A_1, A_2) = A_1$, $\alpha(1, A_1, A_2) = A_2$, and α is a rotation matrix for all $t \in [0, 1]$.

Henceforth, the derivative of α is denoted as $\dot{\alpha}(t) = \frac{\partial}{\partial u}|_{u=t}\alpha(t)$, and the derivative of Exp_A at $H \in T_A SO(n)$ is represented as $(d\text{Exp}_A)_H$ and computed by

$$(d\text{Exp}_A)_{H_1}(H_2) = \frac{d}{dt}|_{t=0} \text{Exp}_A(H_1 + tH_2), \quad \text{for all } H_2 \in T_A SO(n) \simeq T_{H_1}(T_A SO(n)).$$

For two rotation matrices $A_1, A_2 \in SO(n)$ the geodesic distance is given by

$$d(A_1, A_2) = \|\text{Log}_{A_1}(A_2)\| = \|\log(A_1^T A_2)\|. \quad (4.13)$$

where $\|.\|$ corresponds to the Frobenius norm. For each $A_1 \in SO(n)$, there exists a symmetry

$$\varphi_{A_1} : SO(n) \longrightarrow SO(n), \quad A_2 \longrightarrow A_1 A_2^T A_1, \quad (4.14)$$

that reverses geodesics through A_1. φ_{A_1} is an isometry, resulting in the endowment of $SO(n)$ with the structure of a Riemannian symmetric space.

Lemma 4.2 *For $A_1, A_2 \in SO(n)$, let $(d\varphi_{A_1})(A_2)$ denote the derivative of the geodesic symmetry φ_{A_1} at A_2. The following properties are satisfied.*

1. $(d\varphi_{A_1})_{A_2}^{-1} = (d\varphi_{A_1})_{\varphi_{A_1}(A_2)}$, *for all $A_2 \in SO(n)$.*
2. $(d\varphi_{A_1})_{\text{Exp}_{A_1}(H)} \circ (d\text{Exp}_{A_1})_H = -(d\text{Exp}_{A_1})_{-H}$, *for all $H \in T_{A_1} SO(n)$.*

Table 4.1 Geometric toolbox for the special orthogonal group $SO(n)$

Set	$SO(n) = \{A \in \mathcal{M}_n(\mathbb{R}) \mid A^T A = I, \det(A) = 1\}$
Tangent spaces	$T_A SO(n) = \{H \in \mathcal{M}_n(\mathbb{R}) : AH^T + HA^T = 0\}$
Inner product	$\langle H_1, H_2 \rangle_A = \text{trace}(H_1^T H_2)$
Geodesic distance	$d(A_1, A_2) = \|\text{Log}_{A_1}(A_2)\| = \|\log(A_1^T A_2)\|$
Shortest geodesic	$\alpha(t, A_1, A_2) = A_1 \exp(t \log(A_1^T A_2))$
Exponential	$\text{Exp}_A(H) = A \exp(A^T H)$
Logarithm	$\text{Log}_{A_1}(A_2) = A_1 \log(A_1^T A_2)$
Geodesic symmetry	$\varphi_{A_1} : SO(n) \longrightarrow SO(n), \ A_2 \longrightarrow A_1 A_2^T A_1$

The geometric toolbox presented in Table 4.1 summarizes different geometric structures on $SO(n)$.

4.3 Bézier Splines on $SO(n)$

4.3.1 Bézier Curves on $SO(n)$

Given $(k+1)$ points $B_0, B_1, ..., B_k$ on $SO(n)$, the geodesic curve joining the points B_k at $t = 0$ and B_{k+1} at $t = 1$ is given by:

$$\alpha(t; B_k, B_{k+1}) = \exp(t \log(B_k^T B_{k+1}))B_k, \quad t \in [0, 1].$$

The steps involved in constructing a Bézier curve on $SO(n)$ using the De Casteljau algorithm are as follows. Initially, set $B_i^0 = B_i$ for $i = 0, ..., k$. Then, for $i = 0, ..., k - j$ and $j = 1, ..., k$, calculate the points B_i^j using the formula:

$$B_i^j = \exp(t \log((B_i^{j-1})^T B_{i+1}^{j-1}))B_i^{j-1}, \quad t \in [0, 1], \quad (4.15)$$

The last point computed in this algorithm represents a point on the Bézier curve β_k:

$$B_0^k = \beta_k(t; B_0, ..., B_k) \quad (4.16)$$

Example 4.2 (*Cubic Bézier curve on $SO(n)$*) Let B_0, B_1, B_2, B_3 be a rotation matrices on $SO(n)$. Fix $t \in [0, 1]$, for $i = 0, 1, 2, 3$, set $B_0^0 = B_0$, $B_1^0 = B_1$, $B_2^0 = B_2$ and $B_3^0 = B_3$ and consider the shortest geodesic

$$\alpha(t, B_i^{j-1}, B_{i+1}^{j-1}) = \exp(t \log((B_i^{j-1})^T B_{i+1}^{j-1}))B_i^{j-1}, \quad t \in [0, 1], \quad (4.17)$$

4.3 Bézier Splines on $SO(n)$

for $i = 0$ and $j = 1$, we compute points B_0^1, B_1^1, B_2^1 as follows:

$$B_0^1 = \alpha(t, B_0^0, B_1^0) = \exp(t \log(B_0^T B_1)) B_0$$
$$B_1^1 = \alpha(t, B_1^0, B_2^0) = \exp(t \log(B_1^T B_2)) B_1$$
$$B_2^1 = \alpha(t, B_2^0, B_3^0) = \exp(t \log(B_2^T B_3)) B_2$$

Now, as i takes values of 0 and 1, with $j = 2$, we calculate the points B_0^2 and B_1^2 as follows:

$$B_0^2 = \alpha(t, B_0^1, B_1^1) = \exp(t \log((B_0^1)^T B_1^1)) B_0^1$$
$$B_1^2 = \alpha(t, B_1^1, B_2^1) = \exp(t \log((B_1^1)^T B_2^1)) B_1^1$$

Finally, the point

$$B_0^3 = \alpha(t, B_0^2, B_1^2) = \exp(t \log((B_0^2)^T B_1^2)) B_0^2$$

is the point on the cubic Bézier curve β_3 for the fixed $t \in [0, 1]$.

Proposition 4.1 *The Bézier curve $t \to \beta_k(t; B_0, ..., B_k)$ of order k satisfies the following boundary conditions:*

1. $\beta_k(0; B_0, ..., B_k) = B_0$ and $\beta_k(1; B_0, ..., B_k) = B_k$.
2. $\beta_k(t; B_0, ..., B_k) = \beta_k(1 - t; B_k, ..., B_0)$, $t \in [0, 1]$.

Let's denote $\dot{\beta}_k(t; B_0, ..., B_k) = \frac{d}{du}\big|_{u=t} \beta_k(u; B_0, ..., B_k)$ and $\frac{D}{du}\big|_{u=t} \beta_k(u; B_0, ..., B_k)$ as the velocity and covariant acceleration of the Bézier curve β_k of order k, where $t \in [0, 1]$.

Proposition 4.2 *The derivatives of the Bézier curve $t \to \beta_k(t; B_0, ..., B_k)$ of order k satisfy:*

1. $\dot{\beta}_k(0; B_0, ..., B_k) = k\dot{\alpha}(0, B_0, B_1) = k B_0 \log(B_0^T B_1)$.
2. $\dot{\beta}_k(1; B_0, ..., B_k) = k\dot{\alpha}(1, B_{k-1}, B_k) = k B_{k-1} \log(B_{k-1}^T B_k)$.

Proof See [20] for the proof. □

Theorem 4.1 *Let $t \longrightarrow \beta_k(t; B_0, ..., B_k)$ be the Bézier curve of order k on $SO(n)$ with a number of control points B_k for $i = 0, ..., k$. Then, $\beta_k(t; B_0, ..., B_k)$ satisfies:*

(i) $\frac{D}{dt}\big|_{t=0} \dot{\beta}_k(t; B_0, ..., B_k) = k(k - 1)\Omega_0$, where

$$\Omega_0 := \begin{cases} \dot{\alpha}(0, B_1, B_2), & \text{if } B_0 = B_1, \\ (dExp_{B_0})^{-1}_{\dot{\alpha}(0, B_0, B_1)} (\dot{\alpha}(0, B_1, B_2) - \dot{\alpha}(1, B_0, B_1)), & \text{if } B_0 \neq B_1. \end{cases}$$

(ii) $\frac{D}{dt}|_{t=1} \dot{\beta}_k(t; B_0, ..., B_k) = k(k-1)\Omega_k$, where

$$\Omega_k := \begin{cases} -\dot{\alpha}(0, B_{k-2}, B_{k-1}), & \text{if } B_{k-1} = B_k, \\ (dExp_{B_k})^{-1}_{-\dot{\alpha}(1, B_{k-1}, B_k)} (\dot{\alpha}(0, B_{k-1}, B_k) - \dot{\alpha}(1, B_{k-2}, B_{k-1})), & \text{if } B_{k-1} \neq B_k. \end{cases}$$

Proof See [16] for the proof. □

4.3.2 Bézier Splines and Continuity Conditions

We will now establish necessary and sufficient conditions for a Bézier spline on $SO(n)$ to satisfy C^m continuity, where $m = 0, 1, 2$. Let's begin by providing the definition of a Bézier spline on $SO(n)$.

Definition 4.1 Let $0 = t_0 < t_1 < \cdots < t_N = N$ be a distinct and ordered instants of time, and let $B_0^i, ..., B_k^i$ be a set of control points in $SO(n)$, $i = 0, .., N-1$ and $k \in \mathbb{N}$. A composite Bézier curve, also referred to as a picewise Bézier curve or a Bézier spline, formed by N Bézier curves of order k, is given by

$$\beta : [0, N] \to SO(n), \quad t \to \beta_k^i(t - i, B_0^0, ..., B_k^i), t \in [i, i+1], i = 0, ..., N-1. \tag{4.18}$$

The composite Bézier spline, as defined above, is C^∞ on the interval $]t_i, t_{i+1}[$ and is continuous if and only if $B_{k^i}^i = B_0^{i+1}$. This condition is a consequence of the endpoint interpolation property for Bézier curves on $SO(n)$. To attain C^1 continuity, it is crucial to meet the differentiability condition at the joint points, $B_{k^i}^i = B_0^{i+1}$.

Proposition 4.3 *The Bézier spline $\beta : [0, N] \to SO(n)$ defined by Eq. 4.18 is C^1 if the following conditions hold*

$$B_{k^i}^i = B_0^{i+1} \tag{4.19}$$

$$k^i \dot{\alpha}(1, B_{k^i-1}^i, B_{k^i}^i) = k^{i+1} \dot{\alpha}(0, B_0^{i+1}, B_1^{i+1}) \tag{4.20}$$

Proof The proof follows in the same way as the proof of Proposition 3.3. □

Example 4.3 Consider a cubic Bézier spline formed by two cubic Bézier curves β_3^i on $SO(n)$. The cubic Bézier spline is defined as

$$\beta(t) = \begin{cases} \beta_3(t; B_0^0, B_1^0, B_2^0, B_3^0), & \text{if } t \in [0, 1] \\ \beta_3(t - 1; B_0^1, B_1^1, B_2^1, B_3^1), & \text{if } t \in [1, 2] \end{cases}$$

Let's assume that β is continuous, this gives $B_3^0 = B_0^1$. Then, β is C^1 if and only if $\dot{\alpha}(1, B_2^0, B_3^0) = \dot{\alpha}(0, B_0^1, B_1^1)$. By Proposition 4.2, we get:

$$B_2^0 \log((B_2^0)^T B_3^0) = B_0^1 \log((B_0^1)^T B_1^1) \tag{4.21}$$

4.4 Solving Interpolation Problem on $SO(n)$

Proposition 4.4 *The Bézier spline $\beta : [0, N] \to SO(n)$, as defined by Eq. 4.18, exhibits C^2 continuity if the following conditions are satisfied:*

$$B_{k^i}^i = B_0^{i+1} \tag{4.22}$$

$$k^i \dot{\alpha}(1, B_{k^i-1}^i, B_{k^i}^i) = k^{i+1} \dot{\alpha}(0, B_0^{i+1}, B_1^{i+1}) \tag{4.23}$$

$$k^i(k^i - 1)\Omega_i = k^{i+1}(k^{i+1} - 1)\Omega_0 \tag{4.24}$$

Here, Ω_i and Ω_0 are defined in Theorem 4.1.

Proof The proof follows a similar approach to the proof of Proposition 3.4. □

Remark 4.1 As $SO(n)$ is a symmetric Riemannian manifold, as elaborated in [16], the symmetry property can be leveraged to establish the condition under which a Bézier spline achieves C^2 continuity. Let $\beta : [0, N] \to SO(n)$ be a Bézier spline on $SO(n)$ defined by a sequence of N Bézier curves of order $k \geq 2$, denoted as β_k^i, with control points B_k^i for $i = 0, ..., N - 1$. β attains C^2 continuity if the following conditions are satisfied:

$$B_{k^i}^i = B_0^{i+1} \tag{4.25}$$

$$B_1^{i+1} = \varphi_{B_k^i}(B_{k-1}^i) \tag{4.26}$$

$$(d\varphi)_{B_1^{i+1}}(\dot{\alpha}(0, B_1^{i+1}, B_2^{i+1})) = \dot{\alpha}(1, B_{k-1}^i, B_{k-1}^i) - 2\dot{\alpha}(0, B_{k-1}^i, B_k^i) \tag{4.27}$$

where φ is the geodesic symmetry on $SO(n)$ given by Eq. 4.14.

Algorithm 5 can be employed to generate a C^2 Bézier spline on $SO(n)$.

4.4 Solving Interpolation Problem on $SO(n)$

In this section, we aim to extend the methodology, originally designed for constructing a C^2 Bézier spline in Euclidean space, to the Riemannian manifold $SO(n)$. Specifically, given $A_0, ..., A_N$ as a set of $(N + 1)$ distinct points in $SO(n)$ and $0 = t_0 < t_1 < \cdots < t_N = N$ an increasing sequence of time instants, we introduce a conceptually simple framework to create a C^2 Bézier spline $\gamma : [0, N] \to SO(n)$. This curve satisfies the conditions $\gamma(t_k) = A_k$, ; $k = 0, ..., N$. While the generalization of the introduced approach to most Riemannian manifolds is not straightforward, the case of the Lie group $SO(n)$, being a symmetric space with closed-form expressions for essential geometric functions, allows for a comprehensive solution to the problem of constructing a C^2 Bézier spline interpolating a given set of points within this space.

In constructing the Bézier spline γ, we assume its definition involves a sequence of N Bézier curves γ_k^i, where $k \in \{2, 3\}$ and $i = 0, \ldots, N - 1$. In particular, we designate the segments connecting A_0 and A_1 and the segment linking A_{N-1} and

A_N as Bézier curves of order two. All other segments represent Bézier curves of order three. Furthermore, denote $(\widehat{B}_i^-, \widehat{B}_i^+)$ as control points on the left and right-hand side of the interpolation point A_i for $i = 1, \ldots, (N-1)$. The Bézier spline $\gamma : [0, N] \longrightarrow SO(n)$ is then expressed as follows:

$$\gamma(t) = \begin{cases} \gamma_2^0(t; A_0, \widehat{B}_1^-, A_1), & 0 \leq t \leq 1, \\ \gamma_3^i(t - i; A_i, \widehat{B}_i^+, \widehat{B}_{i+1}^-, A_{i+1}), & i - 1 \leq t \leq i, \\ \gamma_2^{n-1}(t - (N-1); A_{N-1}, \widehat{B}_{N-1}^+, A_N), & N - 1 \leq t \leq N. \end{cases}$$

Through Proposition 4.1, it becomes evident that the Bézier spline γ guarantees the interpolation of the first and last control points of each Bézier curve γ_j^i, where $j \in \{2, 3\}$ and $0 \leq i \leq N - 1$. As a result, the continuity of γ at the junction points is adequately ensured. However, the objective is to determine the optimal control points $(\widehat{B}_i^-, \widehat{B}_i^+)$ such that γ belongs to the class C^2. These optimal control points are solutions to the minimization problem, seeking to minimize the mean square acceleration of γ. More precisely, γ is the solution to the minimization problem:

$$\min_{\gamma \in C^2([0,N],SO(n))} E(\gamma) = \int_0^N \left\langle \frac{D^2\gamma(t)}{Dt^2}, \frac{D^2\gamma(t)}{Dt^2} \right\rangle dt \qquad (4.28)$$

4.4.1 Proposed Interpolating Bézier Spline on $SO(n)$

In this section, we delve into the construction of the C^2 interpolating Bézier spline γ on $SO(n)$ that minimizes the cost function given by Eq. 4.28. The step-by-step procedure for constructing γ is outlined below:

1. **Step 1:** For each $i = 1, \ldots, N - 1$, map the set of data points $A_0, \ldots, A_N$ into the tangent space $T_{A_i}SO(n)$.
   Given a sequence of $(N + 1)$ data points $A_0, \ldots, A_N$ in $SO(n)$ with A_j belonging to the cut locus of A_i, we use the Riemannian logarithmic map, expressed by Eq. 4.8, to project the data points $A_0, \ldots, A_N$ into each tangent space $T_{A_i}SO(n)$, where $i = 1, \ldots, N - 1$. The transformed data are then represented by $R^i = (R_0^i, \ldots, R_N^i)$, where $R_k^i = \text{Log}_{A_i}(A_k)$ for $k = 0, \ldots, N$.

2. **Step 2:** Solve the optimization problem Eq. 4.28 on each tangent space $T_{A_i}SO(n)$, $i = 1, \ldots, N - 1$.
   Let $\beta_i : [t_0, t_n] \rightarrow T_{A_i}SO(n)$ denote the Bézier spline on $T_{A_i}SO(n)$, where $i = 1, \ldots, N - 1$. It is defined by N Bézier curves β_j^i, where $j \in \{2, 3\}$ and $0 \leq i \leq N - 1$. Precisely, β_i is given by:

$$\beta_i(t) = \begin{cases} \beta_2^0(t; R_0^i, (\widehat{H}_1^i)^-, R_1^i), & 0 \leq t \leq 1, \\ \beta_3^i(t - i; R_i^i, (\widehat{H}_i^i)^+, (\widehat{H}_{i+1}^i)^-, R_{i+1}^i), & i - 1 \leq t \leq i, \\ \beta_2^{N-1}(t - (N-1); R_{N-1}^i, \widehat{H}_{N-1}^+, R_N^i), & N - 1 \leq t \leq N. \end{cases}$$

4.4 Solving Interpolation Problem on $SO(n)$

Similarly to the Euclidean case, the differentiability condition 2.36 allows us to express $(\widehat{H}_k^i)^+$ in terms of $(\widehat{H}_k^i)^-$. Hence, optimization problem Eq. 4.28 is defined as a function of $(\widehat{H}_k^1)^-$ as follows on $T_{A_i}SO(n)$:

$$\min_{(\widehat{H}_1^i)^-,\dots,(\widehat{H}_{N-1}^i)^-} E((\widehat{H}_1^i)^-,\dots,(\widehat{H}_{N-1}^i)^-) := \min_{(\widehat{H}_1^i)^-,\dots,(\widehat{H}_{N-1}^i)^-} \int_0^1 \|(\ddot{\beta}_2^i)^0(t; R_0^i, (\widehat{H}_1^i)^-, R_1^i)\|_2^2$$

$$+ \sum_{i=1}^{N-2} \int_0^1 \|\ddot{\beta}_3^i(t; R_i^i, (\widehat{H}_i^i)^+, (\widehat{H}_{i+1}^i)^-, R_{i+1}^i)\|_2^2$$

$$+ \int_0^1 \|\ddot{\beta}_2^{n-1}(t; R_{n-1}^i, (\widehat{H}_{N-1}^i)^+, R_N^i)\|_2^2. \quad (4.29)$$

The control points of the C^2 interpolating Bézier curve β_i on $T_{A_i}SO(n)$, solution of the optimization problem Eq. 4.29, are determined through the algorithm described below:

Input: $N \geq 3$, $R = [R_0, \dots, R_N]^T$ a matrix of size $n(N+1) \times n$ containing the $(N+1)$ interpolation points on $T_{A_i}SO(n)$

Output: $H_i = [(H_1^1)^-, \dots, (H_{N-1}^1)^-]^T$ a matrix of size $n(N-1) \times n$ containing the control points of the C^2 Bézier curve β_i in $T_{A_i}SO(n)$.

1: Determine $\widehat{H}_i = [(\widehat{H}_1^1)^-, \dots, (\widehat{H}_{N-1}^1)^-]^T$ using Algorithm 3.
2: Set $(H_1^1)^- = (\widehat{H}_1^1)^-$.
3: Set $(H_2^1)^- = \frac{1}{3}R_0^1 - \frac{1}{2}(H_1^1)^- + \frac{8}{3}R_1^1$.
4: **for** $i = 2 : N - 2$ **do**
5: $\quad (H_{i+1}^1)^- = (H_{i-1}^1)^+ + 4R_i^1 - 4(H_i^1)^-$.
6: **end for**

The detailed computation to obtain the control points of the C^2 interpolating Bézier curve β_i on $T_{A_i}SO(n)$, is similar to the Theorem 3.2.

Remark 4.2 Similar to the Euclidean case, the interpolation points on $T_{A_i}SO(n)$ are adjusted, resulting in the new set of $(N+1)$ interpolation points $\tilde{R} = (\tilde{R}_0^1, \dots, \tilde{R}_N^1)$ on $T_{A_i}SO(n)$, which can be expressed as follows:

$$\tilde{R}_k^1 = R_k^1, \text{ for } k = 0, \dots, N-1, \quad (4.30)$$

$$\tilde{R}_N^1 = 2R_{N-1}^1 + 2(H_{N-1}^1)^+ - 6(H_{N-1}^1)^- + 3(H_{N-2}^1)^+, \quad (4.31)$$

3. **Step 3:** Find the control points of the Bézier spline γ on $SO(n)$.
The Riemannian exponential map Exp_{A_i} defined on $SO(n)$ by Eq. 4.7 is now used to transition the control points of the C^2 Bézier curve β_i from the tangent space $T_{A_i}SO(n)$ to $SO(n)$. Specifically, this transition is expressed as:

$$\widehat{B}_i^- = \text{Exp}_{A_i}(\tilde{h}_i), \ i = 1, \dots, N-1, \quad (4.32)$$

where $\tilde{h}_i$ represents the ith row of H_i in $T_{A_i}SO(n)$ for $i = 1, ..., N - 1$. These resulting points on $SO(n)$ serve as the optimal control points for generating a Bézier spline γ on $SO(n)$. Additionally, the Bézier spline γ is C^1 on $SO(n)$, due to the generalization of Eq. 3.51 to $SO(n)$ in the form $\widehat{B}_i^+ = \mathrm{Exp}_{A_i}(\epsilon_i \mathrm{Exp}_{A_i}^{-1}(\widehat{B}_i^-))$, for $i = 1, ..., N - 1$.

4. **Step 4:** Determine the updated control points for the Bézier spline to achieve C^2 continuity.

Let $\tilde{A}_k = \mathrm{Exp}_{A_i}(\tilde{R}_k^i)$ denote the new interpolation points on $SO(n)$, and B_i^- and B_i^+ represent the updated control points on the left and right sides of the interpolation point $\tilde{A}_i$ responsible for generating the C^2 Bézier spline γ on $SO(n)$. Using the inherent symmetries at the interpolation points, we calculate the control points B_i^- and B_i^+ following the algorithms outlined below:

Input: $N \geq 3$, $\tilde{A} = [\tilde{A}_0, ..., \tilde{A}_N]^T$ a matrix of size $n(N + 1) \times n$ containing the $(N + 1)$ interpolation points on $SO(n)$ and $H_1 = [(H_1^1)^-, ..., (H_{N-1}^1)^-]^T$, the matrix of size $n(N - 1) \times n$ containing the control points of the C^2 Bézier spline β_1 in $T_{\tilde{A}_1}SO(n)$.

Output: $B = [B_1^-, ..., B_{N-1}^-]^T$ a matrix of size $n(N - 1) \times n$ containing the control points of the C^2 Bézier spline γ on $SO(n)$.

1: Set $B_1^- = \mathrm{Exp}_{\tilde{A}_1}((H_1^1)^-)$.
2: Set $B_2^- = \mathrm{Exp}_{B_1^+}\left(\frac{1}{3}\left((d\varphi_{\tilde{A}_1})_{B_1^-}\left(\dot{\alpha}(1, \tilde{A}_0, B_1^-)\right) - 4\dot{\alpha}(0, B_1^-, \tilde{A}_1)\right)\right)$.
3: **for** $i = 2 : N - 2$ **do**
4: $\quad B_{i+1}^- = \mathrm{Exp}_{B_i^+}\left(\left((d\varphi_{\tilde{A}_i})_{B_i^-}\left(\dot{\alpha}(1, B_{i-1}^+, B_i^-)\right) - 2\dot{\alpha}(0, B_i^-, \tilde{A}_i)\right)\right)$.
5: **end for**

The procedure for obtaining equations that govern the control points of the C^2 Bézier spline γ on $SO(n)$ is analogous to the approach outlined in Theorem 3.3.

4.4.2 Experiments

In this section, we present various examples of C^2-smooth Bézier interpolating splines for a given set of points on $SO(3)$ at specified time instants. The numerical examples are structured as follows. Initially, we use a 3D object in $\mathbb{R}^3$ for illustration purposes. We focus on interpolating a finite set of rotation matrices A_0, A_1, A_2, and A_3 in $SO(3)$ at specified time instants ($t = 1, 6, 12, 18$). Subsequently, the method described in the previous section is employed to compute the optimal C^2 interpolating Bézier spline on $SO(3)$, which is then applied to rotate the 3D object. The results are visualized in Fig. 4.1. The provided rotations and time instants are displayed in boxes. Through this example, we can easily check that the resulting splines are smooth, including at the interpolation points.

4.4 Solving Interpolation Problem on $SO(n)$

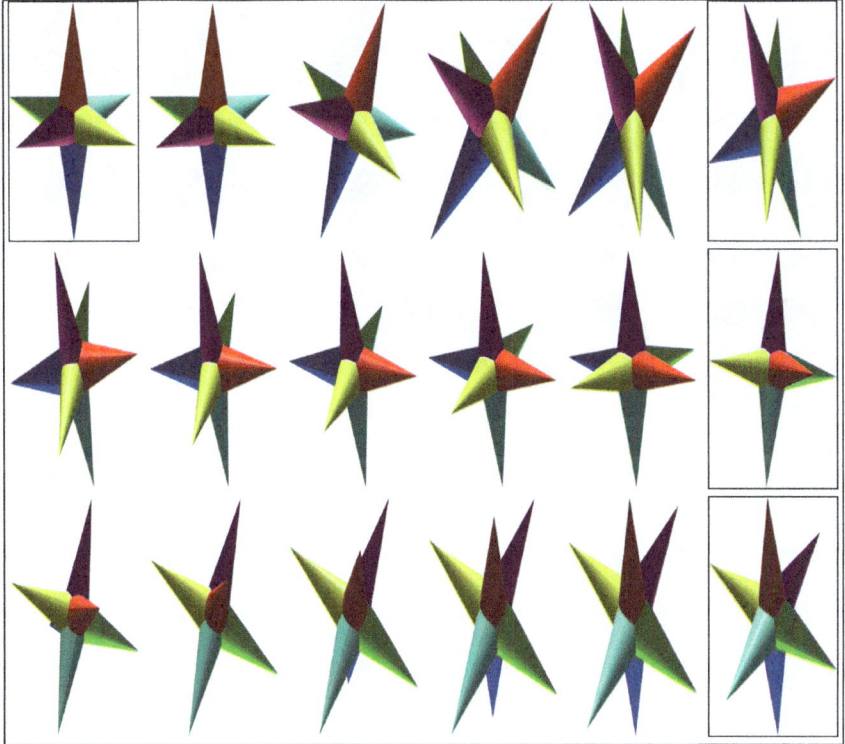

Fig. 4.1 An example of C^2 Bézier curve on $SO(3)$. All rotations are applied to rotate a 3D object for illustration. The initial points are given at time instants ($t = 1, 6, 12, 18$) and are displayed in boxes

As part of potential applications, we would like to point out that the interpolation method developed in this work can be used for more general fitting problems under the assumptions detailed in each algorithm. As a second example, we want to simulate a spline of an observer around a target of interest. For simplicity, we consider an equivalent situation where the observer has a fixed point of view and the target is subject to rotations. Consequently, the problem is reduced to construct an interpolation that passes by some rotation matrices A_0, A_1, A_2, and A_3 in $SO(3)$ at fixed time instants ($t = 1, 5, 9, 13$). To better understand this situation, we fix the target (earth here) and simulate the trajectory of an object (a satellite in this example) as an interpolation on $\mathbb{R}^3 \times SO(3)$. The object has to perform a smooth displacement and rotations at given time instants. We illustrate this idea with a sketch of trajectory in Fig. 4.2.

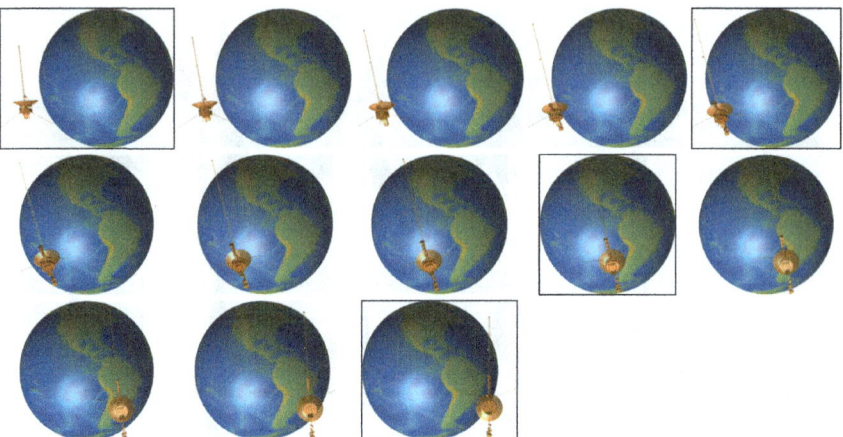

Fig. 4.2 The interpolating path constructed to illustrate a sketch trajectory of 3D object around a sphere. The initial points are given at time instants ($t = 1, 5, 9, 13$) and are displayed in boxes

References

1. Zhao, X. and Liu, Y. and Xie, Z.: The application of acceleration-level quaternion interpolation in the visual-servoing process, 13th International Conference on Computer and Automation Engineering (ICCAE), 2021.
2. Legnani, G, and Fassi, I. and Tasora, A.:A practical algorithm for smooth interpolation between different angular positions Mechanism and Machine Theory, 2021.
3. Haarbach A. and Birdal, T. and Ilic, S.: Survey of higher order rigid body motion interpolation methods for keyframe animation and continuous-time trajectory estimation, International Conference on 3D Vision (3DV), 2018.
4. Sabatini, A.M.: Quaternion based attitude estimation algorithm applied to signals from body-mounted gyroscopes, Electron. Lett, 40(10), 584–586, 2004.
5. Niu, X. and Wang, T.: C^2-continuous orientation trajectory planning for robot based on spline quaternion curve, Assemb. Autom, 38(3), 282–290, 2018.
6. Assa, A. and Janabi-Sharifi, F.: Virtual visual servoing for multicamera pose estimation, IEEE/ASME Trans. Mech, 20(2), 789–798, (2015).
7. Shoemake, K.: Animating rotation with quaternion curves, ACMSIGGRAPH', 85(19), 245-254, (1985).
8. Barr, A.H and Currin, B. and Gabriel, S. and Hughes, J.F: Smooth interpolation of orientations with angular velocity constraints using quaternions, Comput. Graph., 25(2), 313-320, 1985.
9. Hart, J.C and Francis, G.K and Kaufman, L.H: Visualizing quaternion rotation, ACM Trans. Graph, 13(3), 256–276, 1994.
10. Ge, Q.J and Ravani, B.: Geometric construction of Bézier motions, ASME J. Mech.Des, 116, 749–755, 1994.
11. Nielson, G. and Heiland, R.: Animated rotations using quaternions and splines on a 4D sphere, Program. Comput.Software, 18(4), 145–154, 1992.
12. Rodrigues, R.-C. and Silva Leite, F. and Jakubiak, J.: A New Geometric Algorithm to Generate Smooth Interpolating Curves on Riemannian Manifolds, LMS Journal of Computation and Mathematics, 8, 251-266, 2011.
13. Wu, J. and Liu M. and Ding, J. and Deng, M.: Robust rotation Interpolation based on $SO(n)$ geodesic distance, Lecture Notes in Computer Science, 2019.

References

14. Crnkić, A. and Kapić, Z.: Interpolation on the special orthogonal group with highdimensional Kuramoto model, IOP Conf. Series: Materials Science and Engineering 1208, 2021.
15. Kapic, Z. and Crnkic, A.: Interpolating rotations with non-Abelian Kuramoto model on the 3-sphere Advanced Technologies, Systems, and Applications VI, Proceedings of the International Symposium on Innovative and Interdisciplinary Applications of Advanced Technologies (IAT 2021), 2021.
16. Popiel, T. and Noakes, L.: Bézier curves and C^2 interpolation in Riemannian manifolds, J. Approx. Theory, 148(2), 111–127, 2007.
17. Geir, B. and Klas, M. and Olivier, V.:Numerical Algorithm for C^2-splines on Symmetric Spaces, SIAM Journal on Numerical Analysis, 56 (4), 2623–2647, 2018.
18. Sigurdur H.: Differential Geometry, Lie Groups, and Symmetric spaces, Academic Press, New York, 1978.
19. Frank Adams, J.:Lectures on Lie Groups, the University of Chicago Press, Midway Reprints Series, 1983.
20. Crouch, P., Kun, G. and Silva Leite, F.: The de Casteljau algorithm on Lie groups and spheres, J. Dyn. Control Syst. 5, 397–429, 1999.

# Chapter 5
Spline Interpolation on Stiefel and Grassmann Manifolds

In various real-world applications, the Stiefel and Grassmann manifolds are commonly employed for representation within Riemannian manifolds [1–3]. However, a persistent challenge in many of these applications stems from the intricate geometric structures inherent in these manifolds [4]. As real-world applications increasingly involve non-vector data, numerous algorithms for manifold embedding and manifold learning have been introduced to address these challenges. Recent efforts in this direction have focused on the development of essential geometric and statistical tools, including the Riemannian exponential map and its inverse, means, distributions, and geodesics [5–7].

Motivated by the success of these methods, this chapter addresses the challenge of achieving smooth interpolation for a given dataset on Grassmann and Stiefel manifolds. The proposed approach extends the De Casteljau algorithm, transforming the interpolation problem into the well-established Euclidean framework. This adaptation allows us to leverage the extensive toolkit available for spline interpolation directly. The resulting interpolated spline exhibits several desirable characteristics, including the existence and optimality of the solution in various common scenarios. Additionally, we explore the specific structures related to chosen Riemannian metrics, offering additional computational benefits for practical applications.

5.1 Problem Formulation

We are addressing the challenge of fitting smooth curves to a finite set of data points on a special class of Riemannian manifolds, specifically the Grassmann manifold, encompassing all p-dimensional subspaces of $\mathbb{R}^n$, and the Stiefel manifold, consisting of p-orthonormal vectors in $\mathbb{R}^n$. More precisely, when provided with a finite set of points $X_0, X_1, \ldots, X_N$ on $\mathcal{M}$ and distinct, ordered time instants

($0 = t_0 < t_1 < ... < t_N = N$), the objective is to find a spline $\alpha : [0, N] \to \mathcal{M}$ that optimally fits the given data $X_0, X_1, \ldots, X_N$ while ensuring sufficient smoothness. Notably, the focus is on the search space of smooth regression splines, where the data points adhere to orthogonality constraints, and t_i represents distinct and ordered time instants. This problem is prevalent in real-world scenarios, necessitating specific optimization methods commonly referred to as optimization on Riemannian manifolds [4, 8]. To strike a balance between achieving the desired smoothness and closely approximating the data points at the given t_i, one approach is to formulate the spline α as the minimizer of the following functional:

$$E : \Gamma \to \mathbb{R}^+$$
$$\alpha \mapsto \frac{\lambda}{2} \int_{t_0}^{t_N} <\frac{D^2\alpha(t)}{Dt^2}, \frac{D^2\alpha(t)}{Dt^2}>_\mathcal{M} dt + \frac{1}{2N} \sum_{i=0}^{N} d_\mathcal{M}^2(\alpha(t_i), X_i) \quad (5.1)$$

where Γ denotes the underlying space of admissible C^2 splines in $\mathcal{M}$, $<.,.>_\mathcal{M}$ and $d_\mathcal{M}$ denote the Riemannian metric and distance, respectively.

Riemannian optimization on Stiefel and Grassmann manifolds is generally challenging due to the geometric structure of $\mathcal{M}$ and the orthogonality constraints characterizing these manifolds. Consequently, significant efforts have been invested in developing crucial geometric and statistical tools, such as the Riemannian exponential map and its inverse, means, distributions, geodesic arcs, etc [5]. For example, Batzies et al.[9] provided a simple closed-form expression for the geodesic arc joining two points on Grassmann manifolds based solely on the given data points. Hüper et al.[10] proposed a scheme to compute the Karcher mean on Grassmann manifolds. However, relatively little has been done in this direction for Stiefel manifolds. Edelman et al. [8] derived an accurate expression for geodesics on Stiefel manifolds starting at a given point with a specified direction. Furthermore, they proposed numerical methods to compute exponential maps with both Euclidean and canonical metrics. Nevertheless, explicit expressions for geodesics that join two points on Stiefel manifolds are not yet known.

To tackle this challenge, in [11], the concept of a "quasi-geodesic" is introduced- a curve with constant speed, constant covariant acceleration, and constant geodesic curvature that connects two points on the Stiefel manifold. In [12, 13], the problem is addressed by computing the Riemannian log map equipped with the canonical metric. It's worth noting that methods for computing the Riemannian log map with the Euclidean metric have been discussed in [14, 15]. By leveraging these explicit formulas, one can solve other crucial problems, including fitting and interpolation of data points on Grassmann and Stiefel manifolds.

In this context, Hüper et al.[16] proposed a method for generating interpolating curves on Grassmann manifolds based on the rolling technique. Batzies et al. [9] derived the necessary condition for the geodesic that minimizes the least-squares problem on Grassmann manifolds. Another approach, described by Rentmeesters [12], utilizes the Jacobi field approach to approximate a given set of data points on the Grassmannian using the gradient descent technique. Additionally, the recent paper [17] introduces a new method for generating smooth interpolating curves on Stiefel manifolds. This method involves substituting geodesics in the geometric De Casteljau algorithm on manifolds with a successive quasi-geodesic interpolation. Hong et al. [18] proposed an intrinsic geodesic regression model, generalizing classical linear least-squares regression on the Grassmannian.

In this chapter, we present a geometric algorithm for generating solutions to problem 5.1, expressed as interpolating Bézier splines with a specified degree of smoothness. The focus is on the Grassmann manifold, chosen as an illustrative example of symmetric spaces. Leveraging the definition of geodesic curves and capitalizing on the rich and elegant structure of these manifolds, we introduce a novel approach to construct a C^2 Bézier spline that interpolates a given dataset of points at prescribed knot times.

For the Stiefel manifold, the goal is analogous, involving regression on a homogeneous space [4]. The objective is to estimate or predict missing data from a limited set of available observations, where observations represent data points obtained through temporal acquisitions. Specifically, in this context, we focus exclusively on a continuous-time model of class C^1, addressing the problem of regularized non-linear regression from finite observations [19].

In both cases, we start with the energy minimization formulation of linear least-squares in the Euclidean space $\mathbb{R}^n$ and extend this concept to the considered manifolds. The proposed method is geometrically straightforward, extensible, and easy to implement. We substantiate the effectiveness of the proposed approach through various illustrative experiments.

5.2 Geometry of the Stiefel Manifold

The compact Stiefel manifold $\mathcal{M} = St(n, p)$ is a matrix manifold of p-dimensional orthonormal frames in $\mathbb{R}^n$. As an embedded submanifold of $\mathbb{R}^{n \times p}$, $\mathcal{M}$ is defined by

$$\mathcal{M} = \{X \in \mathbb{R}^{n \times p} \mid X^T X = I_p\}.$$

where I_p denotes the $p \times p$ identity matrix. When $p = 1$, we simply have the unit sphere S^{n-1}, while when $p = n$, we have the orthogonal Lie group $O(n)$ and when $p = n - 1$, we obtain the special orthogonal group $SO(n)$.

Remark 5.1 The Stiefel manifold can also be viewed as a quotient manifold of the orthogonal Lie group $O(n)$. In fact, two points Q_0 and Q_1 in the orthogonal Lie group $O(n)$ represent the same point in $\mathcal{M}$ if

$$Q_1 = Q_0 \begin{pmatrix} I_p & 0 \\ 0 & P \end{pmatrix},$$

for $P \in O(n-p)$. The linear left action of $O(n)$ in $\mathcal{M}$ is transitive (i-e, for every pair of elements X_1 and X_2, there is an element $Q \in O(n)$ such that $QX_1 = X_2$). Besides, the isotropy group at the identity frame $X = \begin{bmatrix} I_p \\ 0 \end{bmatrix} \in St(n,p)$ is isomorphic to $O(n-p)$. We remind that the isotropy group the identity frame X is a subgroup of $O(n)$ fixing X, i-e $\forall Q \in O(n)$, $QX = X$. Consequently, $\mathcal{M}$ is diffeomorphic to $O(n)/O(n-p)$ which turns the matrix manifold into a homogeneous space.

For any matrix representative $X \in \mathcal{M}$, the tangent space of $\mathcal{M}$ at X is defined as

$$T_X \mathcal{M} = \{Z \in \mathbb{R}^{n \times p} \mid X^T Z + Z^T X = 0\}.$$

Therefore, both the dimensions of $T_X \mathcal{M}$ and $\mathcal{M}$ are $np - \frac{1}{2}p(p+1)$. The Stiefel manifold can be equipped with different Riemannian metrics, such as the Euclidean metric and the canonical metric. In the special cases when $p = 1$ and $p = n$, these two metrics coincide. However, for other values of p, they differ and lead to distinct formulas for geodesics and parallel translation. For the purposes of this chapter, we choose to endow $\mathcal{M}$ with the canonical metric. Specifically, for $X \in \mathcal{M}$ and $Z_1, Z_2 \in T_X \mathcal{M}$, the canonical metric on $T_X \mathcal{M}$ is defined by

$$\langle Z_1, Z_2 \rangle_X = \operatorname{trace}(Z_1^T (I_n - \frac{1}{2} X X^T) Z_2). \tag{5.2}$$

Geodesics on a Riemannian manifold represent locally shortest curves parametrized by arc length. For a given curve $\gamma : [0,1] \to \mathcal{M}$, these geodesics satisfy the following second-order differential equation:

$$\ddot{\gamma} + \dot{\gamma}\dot{\gamma}^t \gamma + \gamma \left((\gamma^t \dot{\gamma})^2 + \dot{\gamma}^t \dot{\gamma} \right) = 0. \tag{5.3}$$

It is evident that solving this equation numerically poses challenges. Fortunately, the canonical structure facilitates a practical decomposition of the tangent space, simplifying the characterization of geodesics.

Proposition 5.1 *Let X be a matrix representation on $\mathcal{M}$ and Z a tangent vector on $T_X \mathcal{M}$. Then the geodesic $\gamma : [0, 1] \to \mathcal{M}$ such that $\gamma(0) = X$ and $\frac{\partial \gamma}{\partial t}|_{t=0} = Z$ is given by*

$$\gamma(t) = XM(t) + QN(t), \tag{5.4}$$

where $M(t)$ and $N(t)$ are p-by-p matrices defined by

$$\begin{bmatrix} M(t) \\ N(t) \end{bmatrix} = \exp\left(t \begin{pmatrix} A & -R^t \\ R & 0 \end{pmatrix} \right) \begin{bmatrix} I_p \\ 0 \end{bmatrix}. \tag{5.5}$$

Proof The proof starts with a decomposition of the tangent vector Z into its horizontal and vertical components with respect to the base point X, $Z = XX^T Z + (I_n - XX^T)Z$. Then, by letting $A = X^T Z$ a skew symmetric matrix and by means of a

QR decomposition of $(I_n - XX^T)Z$, we get $Z = XA + QR$, which establishes the formula of the geodesic. □

Corollary 5.1 *Let $\gamma : [0, 1] \to \mathcal{M}$ be a geodesic such that $\gamma(0) = X_1$ and $\frac{\partial \gamma}{\partial t}(t)|_{t=0} = Z \in T_{X_1}\mathcal{M}$. The Riemannian exponential map $\mathrm{Exp}_{X_1} : T_{X_1}\mathcal{M} \to \mathcal{M}$ that sends a Stiefel tangent vector Z to the endpoint $\gamma(1) = X_2$ is given by:*

$$\mathrm{Exp}_{X_1}(Z) = X_1 M + QN = X_2 \in \mathcal{M}, \tag{5.6}$$

where M and N are the same as described in Eq. 5.5.

Conversely, when given two points X_1 and $X_2 \in \mathcal{M}$, the inverse exponential map $\mathrm{Exp}_{X_1}^{-1}$ (also known as the logarithmic map Log_{X_1}) allows for the recovery of the tangent vector $Z = \mathrm{Log}_{X_1}(X_2)$. Formulas for computing the Riemannian log map on the Stiefel manifold with respect to the Euclidean metric are provided in [14, 15]. As far as we know, there exist two different approaches for evaluating the log map on the Stiefel manifold with respect to the canonical metric [12, 13]. In this chapter, we adopt the method presented in [13], assuming that any two points belong to a geodesic ball with an injectivity radius determined in [12].

Corollary 5.2 *The geodesic arc joining X_1 to X_2 in $\mathcal{M}$ can be parameterized explicitly by:*

$$\gamma(t, X_1, X_2) = \mathrm{Exp}_{X_1}\left(t \mathrm{Log}_{X_1}(X_2)\right), \; t \in [0, 1]. \tag{5.7}$$

5.3 Geometry of the Grassmann Manifold

The real Grassmann manifold $\mathcal{M} = G_{n,p}$ is defined as the set of p-dimensional $\mathbb{R}$-linear subspace of $\mathbb{R}^n$. It is a smooth and compact manifold of dimension $p(n - p)$. A point $\mathfrak{X} \in \mathcal{M}$ is a linear subspace that can be numerically represented as the span of a full-rank n-by-p matrix X:

$$\mathcal{M} = \{\mathfrak{X} = \mathrm{span}(X), X \in \mathbb{R}^{n \times p}, \mathrm{rank}(X) = p\}.$$

Remark 5.2 We can readily verify that the matrix representative $X \in \mathfrak{X} \subset \mathcal{M}$ is a point on the Stiefel manifold $St(n, p)$. Specifically, $\mathcal{M}$ is a quotient space of $St(n, p)$ through the action of the orthogonal Lie group $O(n)$:

$$\mathcal{M} = St(n, p)/O(p).$$

Given $X \in \mathcal{M}$, we identify the tangent space of $\mathcal{M}$ by

$$T_X \mathcal{M} = \{S \in \mathbb{R}^{n \times p} \mid X^T S = 0\}.$$

In contrast to the Stiefel manifold, introducing a metric based on the quotient space structure of $\mathcal{M}$ or one inherited from the Euclidean space $\mathbb{R}^{n \times n}$ on the tangent space $T_X\mathcal{M}$ leads to the same metric. Therefore, we will endow $\mathcal{M}$ with the one induced from the Frobenius inner product on $\mathbb{R}^{n \times n}$, for $S_1, S_2 \in T_X\mathcal{M}$, given by

$$\langle S_1, S_2 \rangle_X = \mathrm{tr}(S_1^T S_2).$$

Proposition 5.2 *Let A be a p-by-p skew-symmetric matrix and B an arbitrary $(n - p)$-by-p matrix. A geodesic $\gamma : [0, 1] \to O(n)$ starting at $P \in O(n)$ in the direction of the tangent vector $P \begin{pmatrix} A & -B^t \\ B & 0 \end{pmatrix}$ lying on the horizontal space at P, is defined by*

$$\gamma(t) = P \exp\left(t \begin{pmatrix} A & -B^t \\ B & 0 \end{pmatrix}\right). \tag{5.8}$$

Then, geodesics in $\mathcal{M}$ are projections of the orthogonal Lie group $O(n)$ geodesics under the Riemannian submersion $\pi_1 : O(n) \to G_{n,p}$, that is a geodesic $\tilde{\gamma} : [0, 1] \to G_{n,p}$ is given by $\tilde{\gamma}(t) = [\gamma(t)]$.

Corollary 5.3 *The exponential map $\mathrm{Exp}_X : T_X\mathcal{M} \to \mathcal{M}$ is defined as*

$$\mathrm{Exp}_X(S) = (XW\cos(\Sigma) + V\sin(\Sigma))W^t, \tag{5.9}$$

where $(V\Sigma W^t)$ is the compact Singular Value Decomposition of the tangent vector $S \in T_X\mathcal{M}$.

Gallivan et al. [20] have proposed a method to compute the inverse of the exponential map which is based on the representation of $\mathcal{M}$ with $SO(n)/(SO(p) \times SO(n - p))$ and some results from linear algebra, the CS (cosine-sine) decomposition in particular.

Proposition 5.3 *Let X_1, and $X_2 \in \mathcal{M}$. The log-map $\mathrm{Log}_{X_1}(X_2) : \mathcal{M} \to T_{X_1}\mathcal{M}$ is defined as,*

$$\mathrm{Log}_{X_1}(X_2) = W_2 \Sigma W_1^t, \tag{5.10}$$

where W_2, W_1 are given by the CS decomposition

$$\begin{bmatrix} X_1^t X_2 \\ (I_n - X_1 X_1^t)X_2 \end{bmatrix} = \begin{bmatrix} W_1 \cos(\Sigma)V^t \\ W_2 \sin(\Sigma)V^t \end{bmatrix}. \tag{5.11}$$

Corollary 5.4 *The shortest geodesic arc joining X_1 to X_2 in $\mathcal{M}$ is given by*

$$\gamma(t, X_1, X_2) = \mathrm{Exp}_{X_1}\left(t\mathrm{Log}_{X_1}(X_2)\right), \; t \in [0, 1]. \tag{5.12}$$

Here and subsequently,

$$\dot{\gamma}(t, X_1, X_2) := \frac{\partial}{\partial u}|_{u=t} \gamma(u, X_1, X_2), \quad (5.13)$$

and $(d\mathrm{Exp}_{X_1})_S$ stands for the derivative of Exp_{X_1} at $S \in T_{X_1}\mathcal{M}$.

Definition 5.1 For any $X \in \mathcal{M}$, the map $\varphi : \mathcal{M} \to \mathcal{M}$ defined by

$$\varphi(X) = PXP^{-1}, \text{ where } P = \begin{pmatrix} I_p & 0 \\ 0 & -I_{n-p} \end{pmatrix}, \quad (5.14)$$

is an involutive automorphism, wich turns the Grassmann manifold into a Riemannian symmetric space.

5.4 Interpolation Problem on the Stiefel Manifold

In this section, we present a method for constructing a C^1 interpolating Bézier spline designed to smooth data constrained on the Stiefel Manifold $\mathcal{M}$, equipped with its canonical Riemannian metric. Let's consider $(N + 1)$ distinct data points $X_0, ..., X_N$ in $\mathcal{M}$ corresponding to time instants $t_i = i, i = 0, ..., N$. The objective is to estimate a spline $\alpha : [0, N] \to \mathcal{M}$ that minimizes the cost functional 5.1 while satisfying the following properties: (i) $\alpha(t_i) = X_i, i = 0, ..., N$, (ii) α is composed of N Bézier curves of order j, (iii) α is of class C^1.

5.4.1 Bézier spline on the Stiefel Manifold

Definition 5.2 The Bézier curve $\alpha_j : [0, N] \to \mathcal{M}, t \to \alpha_j(t; V_0, ..., V_j)$ of order j parametrized by $(j + 1)$ control points $V_0, ..., V_j$ is defined as follows. Consider the point $V_i^0 = V_i$ and we iterate the construction of further points. In fact, for $i = 0, ..., j - k, k = 1, ..., j$,

$$V_i^k = \alpha_k(t, V_i, ..., V_{i+k}) = \mathrm{Exp}_{V_i^{k-1}}\left(t\mathrm{Log}_{V_i^{k-1}}(V_{i+1}^{k-1})\right), t \in [0, 1]$$

represent the $i$ith point of the kth step of the De Casteljau algorithm, and thus $\alpha_j(t; V_0, ..., V_j) = V_0^j$.

The resulting Bézier spline $\alpha : [0, N] \to \mathcal{M}$ is composed exclusively of a series of N Bézier curves α_j^i, where $j \in \{2, 3\}$ and $i = 0, ..., N - 1$, with the initial and final curves being quadratic Bézier curves and all the intermediate ones being cubic.

Definition 5.3 Assume that there exists two artificial control points $(\widehat{Y}_i^-, \widehat{Y}_i^+)$ on the left and on the right hand side of the interpolation point X_i for $i = 1, ..., (N-1)$ defining a chain of Bézier curve α_j^i, $j \in \{2, 3\}$, $0 \leq i \leq N-1$. The Bézier spline $\alpha : [0, N] \longrightarrow \mathcal{M}$ is then given by

$$\alpha(t) = \begin{cases} \alpha_2^0(t; X_0, \widehat{Y}_1^-, X_1), & 0 \leq t \leq 1, \\ \alpha_3^i(t-i; X_i, \widehat{Y}_i^+, \widehat{Y}_{i+1}^-, X_{i+1}), & i-1 \leq t \leq i \\ \alpha_2^{N-1}(t-(N-1); X_{N-1}, \widehat{Y}_{N-1}^+, X_N), & N-1 \leq t \leq N \end{cases}$$

As the Bézier spline α interpolates the first and last control points of each Bézier curve α_j^i, $j \in \{2, 3\}$, $0 \leq i \leq N - 1$, we are left with the task of determining the remaining control points $(\widehat{Y}_i^-, \widehat{Y}_i^+)$, $i = 0, ..., N - 1$. Note that we want α to be at least C^1. It is immediate, by construction, that the spline α is C^∞ on $]t_i, t_{i+1}[$, for $i = 1, ..N - 1$. We are thus looking to ensure the differentiability condition at the knot points. The main idea to handle this issue is to treat the fitting problem on different tangent space. Specifically, let $X_0, ..., X_N$ be a set of distinct given points in $\mathcal{M}$ with X_l being in the cut locus of X_i, $i \neq l$. The cut locus of X_l in $\mathcal{M}$, in turn, is the set of points in $\mathcal{M}$ where the geodesics starting at X_l stop being length-minimizing. By means of the algorithm of the logarithm map developed in [12, 13] for the Stiefel manifold, we transport data points $X_0, ..., X_N$ to $T_{X_i}\mathcal{M}$ at a point $X_i \in \mathcal{M}$, $i = 1, ..., N - 1$. Let us denote the mapped data by $Z^i = (Z_0^i, ..., Z_N^i)$ with $Z_m^i = \text{Log}_{X_i}(Z_m)$ for $m = 0, ..., N$. Now, the next concern is to search for the control points of a C^1 Bézier spline on $T_{X_i}\mathcal{M}$, $i = 1, ..., N - 1$. From this tangential solution, the Riemannian exponential map Exp_{X_i} defined on $\mathcal{M}$ by Eq. 5.6 will bring back the solution to the matrix manifold $\mathcal{M}$. The resulting Bézier spline α is then reconstructed with De Casteljau algorithm and we prove that is optimal.

5.4.2 C^1 Bézier Spline on the Stiefel Manifold

Let $\beta : [0, N] \to T_{X_i}\mathcal{M}$ represent the Bézier spline on $T_{X_i}\mathcal{M}$, $i = 1, ..., N - 1$, defined in the same way as the Bézier spline on $\mathcal{M}$, consisting of N Bézier curves β_k^i, where $k \in \{2, 3\}$ and $0 \leq i \leq N - 1$. Let $(B_m^i)^-$ and $(B_m^i)^+$ denote the control points on the left and right sides of the interpolation point Z_m, for $m = 1, ..., (N - 1)$. Hence,

$$\beta(t) = \begin{cases} \beta_2^0(t; Z_0^i, (B_1^i)^-, Z_1^i), & 0 \leq t \leq 1, \\ \beta_3^i(t-i; Z_i^i, (B_i^i)^+, (B_{i+1}^i)^-, Z_{i+1}^i), & i-1 \leq t \leq i \\ \beta_2^{N-1}(t-(N-1); Z_{N-1}^i, (B_{N-1}^i)^+, Z_N^i), & N-1 \leq t \leq N \end{cases}$$

5.4 Interpolation Problem on the Stiefel Manifold

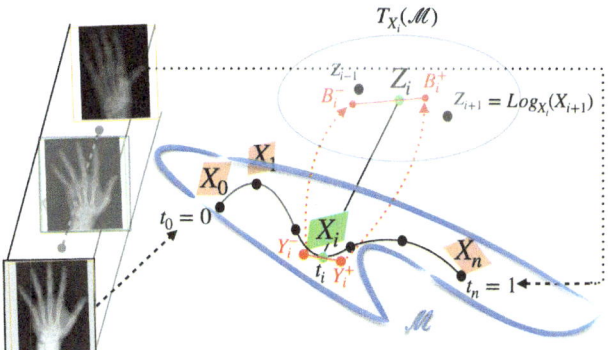

Fig. 5.1 Geometrical illustration of the Riemannian manifold $\mathcal{M}$ and its tangent space $T_{X_i}\mathcal{M}$ at $X_i \in \mathcal{M}$. γ is an interpolating spline with $X_t = \gamma(t)$ for $t \in I$ which verifies $\gamma(t)^T \gamma(t) = I_p$ for all $t \in I$

Proposition 5.4 *The optimization problem Eq. 5.1, which is not easy to solve directly on $\mathcal{M}$ is simplified to an Euclidean cost function on $T_{X_i}\mathcal{M}$, $i = 1, ..., N-1$, given by,*

$$\min_{(B_1^i)^-,...,(B_{N-1}^i)^-} E((B_1^i)^-,...,(B_{N-1}^i)^-) := \min_{(B_1^i)^-,...,(B_{N-1}^i)^-} \int_0^1 \|\ddot{\beta}_2^0(t; Z_0^i, (B_1^i)^-, Z_1^i)\|^2$$
$$+ \sum_{m=1}^{N-2} \int_0^1 \|\ddot{\beta}_3^i(t; Z_m^i, (B_m^i)^+, (B_{m+1}^i)^+, Z_{m+1}^i)\|^2 + \int_0^1 \|\ddot{\beta}_2^{N-1}(t; Z_{N-1}, (B_{N-1}^i)^+, Z_N^i)\|^2$$
(5.15)

where $\|.\|$ represents the canonical norm on the tangent space $T_{X_i}\mathcal{M}$, $i = 1, ..., N$.

The problem 5.1 is now addressed similarly to the Euclidean case $\mathcal{M} = \mathbb{R}^n$. Specifically, we demonstrate that the solutions to the mean square acceleration are precisely the control points of β. Additionally, we provide a geometric illustration of the proposed approach in Fig. 5.1.

Theorem 5.1 *Given $X_0, ..., X_N$ a set of $(N+1)$ data points in $\mathcal{M}$ and $B^i = [(B_1^i)^-, ..., (B_{N-1}^i)^-]^T$ a matrix of size $(n(N-1) \times n)$ containing the $(N-1)$ control points that generate β_i in each tangent space $T_{X_i}\mathcal{M}$, for $i = 1, ..., N-1$. Then, the Bézier spline $\alpha : [0, N] \to \mathcal{M}$ interpolating the data points X_i on $\mathcal{M}$ is of class C^1 and is uniquely defined by the set of control points $Y = [\widehat{Y}_1^-, ..., \widehat{Y}_{N-1}^-]^T \in \mathbb{R}^{n(N-1) \times n}$ where the rows of $\widehat{Y}$ are given by:*

$$\widehat{Y}_i^- = Exp_{X_i}((B_i^i)^-), \ i = 1, ..., N-1,$$
(5.16)

Proof The proof closely parallels that of Theorem 3.2. □

Algorithm 8 synthesizes all steps needed to construct the C^1 solution on $\mathcal{M}$.

Algorithm 8 C^1 Solution on the Stiefel manifold $\mathcal{M}$.

Input: $N \geq 3$, $(X_0, ..., X_N)$ a matrix of size $N(n+1) \times N$ containing the $(N+1)$ interpolation points on $\mathcal{M}$.
Output: $\widehat{Y}$
1: **for** $i = 1 : N-1$ **do**
2:    Compute $Z = [Z_0^i, ..., Z_N^i]^T$ a matrix of size $N(n+1) \times N$ containing the $(N+1)$ interpolation points on $T_{X_i}\mathcal{M}$.
3:    **for** $k = 0 : N$ **do**
4:      $Z_k^i = \text{Log}_{X_i}(X_k)$
5:    Compute $B^i = [(B_1^i)^-, ..., (B_{N-1}^i)^-]^T$ a matrix of size $N(n-1) \times N$ containing the $(N-1)$ control points of the C^1 Bézier curve β_i on $T_{X_i}\mathcal{M}$, using Algorithm 3.
6:    Compute control point $\widehat{Y}_i^- = \text{Exp}_{X_i}((\widehat{B}_i^i)^-)$ on $\mathcal{M}$.
7:    **end for**
8: **end for**
9: **return** $\widehat{Y}$

5.4.3 Experiments

We consider numerical and real-world examples where data points $X_0, X_1, \ldots, X_N$ on a manifold $\mathcal{M}$ are not necessarily parametric. So we can not restrict applications to a class of X_i. We display some examples of random points on a Stiefel manifold $\mathcal{M}$ in Fig. 5.2. For each point X_i, we plot the components $(X_i[1])$, $(X_i[2])$ and $(X_i[3])$ in blue, green and red, respectively. The coordinates are then connected with edges to show the shape of the triangle. This illustration outlines how the positions (components) change and how the shape of the triangle evolves from one point to another. We also note that those points are random inside a geodesic ball so that all the properties detailed in the previous sections are satisfied in order to define an interpolating spline between them. Examples are random and different which capture large variability for illustration. Moreover, Fig. 5.3 shows examples from real-world data. We can observe that the numerical examples, even random, provide good candidates for tests.

In the rest of this section, we illustrate the performance of the proposed method via different experiments. In all cases, we have considered a finite set of data points $X_0, X_1, \ldots, X_N$ on a Stiefel manifold $\mathcal{M}$. Each data point X_i is given at a fixed time instant t_i with $i = 0, 1, \ldots, N$ and $(t_0, t_N) = (0, 1)$, for simplicity. In the first setting, we consider a very common situation where data points are elements on $St(n, 1) = S^{n-1}$. This situation is standard in many applications. To cite but a few popular ones: Normalized directions, longitudinal data, and rotations which is equivalent to $SO(n)$. In the second regression setting, we consider another example where data points are elements on $St(n, p)$ where $p = 2$ and $n = 3$.

5.4 Interpolation Problem on the Stiefel Manifold

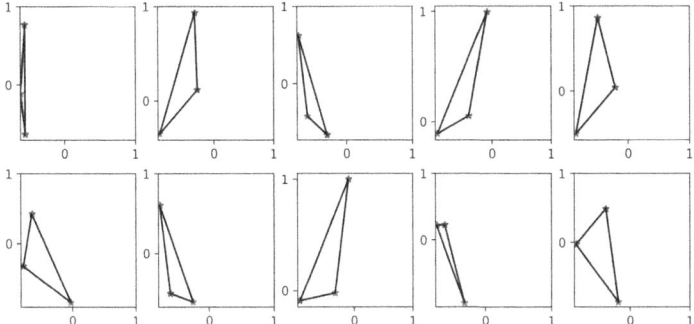

Fig. 5.2 Random numerical examples on $\mathcal{M} = St(3, 2)$

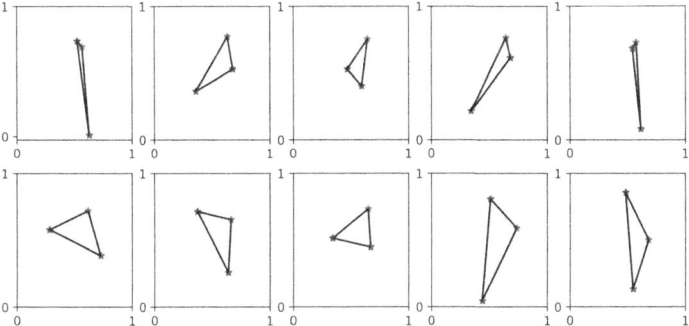

Fig. 5.3 Real data examples on $\mathcal{M} = St(3, 1)$

5.4.3.1 Case 1: Numerical Examples on $St(3, 1)$

These experiments concern the problem where observations are on the finite unit sphere. We consider this case for three reasons: (i) First it is possible to visualize a path on the sphere and (ii) we validate the solution on the simplest example of Stiefel manifolds. Actually, this is a very common problem in many applications, e.g., virtual reality, autonomous driving, and robot navigation [21]. It is very known that geodesics exist and are unique for non antipodal points. Both the Riemannian exponential and its inverse are diffeomorphisms inside a ball of radius π. In all examples, we display the resulting path using the algorithm detailed in the chapter and not a specific formulation for spherical data. We remind that the problem of regression with cubic splines can be efficiently solved in this case. However, the strategy outlined in this chapter is different: We show that the proposed method produces good estimators when the Stiefel manifold coincides with the sphere. Thus, and for visualization purposes, we consider the case $(n, p) = (3, 1)$ and show original data points (black) and the optimal α (red) with different values of N. The time instants are uniformly spaced in [0, 1]. See the three different examples in Fig. 5.4.

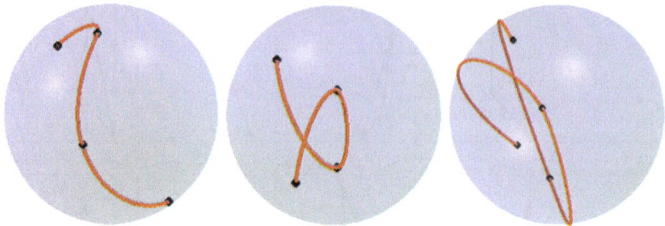

Fig. 5.4 Examples with different data points on $\mathcal{M} = St(3, 1)$

5.4.3.2 Case 2: Numerical Examples on $St(3, 2)$

These experiments concern the nonlinear regression problem where observations are elements of Case $St(3, 2)$. They can be also considered as elements on the unit sphere, modulo rigid transformations, usually denoted Σ_n^p and called the Kendall space. They were largely studied for analyzing biological data [19]. We remind that the main objective is to show that the proposed method is successful in the case of $St(3, 2)$ and remains more general for other cases. Otherwise, geodesics, the exponential map, log map are detailed in [19]. We display the resulting path using the algorithm detailed in the chapter and not a specific formulation for this manifold when equipped with a different structure. The proposed strategy is different: demonstrate that the proposed method produces good estimators when the Stiefel manifold coincides with this manifold. For visualization purposes, we consider the case $(n, p) = (3, 2)$ and show the original data points and the optimal path α with different values of N. The time instants are uniformly spaced in $[0, 1]$. See Fig. 5.5 for an example of α interpolating $(X_0, X_1, X_2, X_3) \in St(3, 2)$ at $t \in \{0, \frac{1}{3}, \frac{2}{3}, 1\}$.

Following the same idea from the previous example in Fig. 5.5, we show another interpolating α using 5 points on $\mathcal{M} = St(3, 2)$ in Fig. 5.6. In the first example (X_1, X_2, X_3) are generated randomly in the ball $(X_0, 0.5\pi)$. In the second example (X_1, X_2, X_3, X_4) are generated randomly in the ball $(X_0, 0.25\pi)$. Considering the shape of the triangles and the coordinates (colored points) we can see that points in the first example are distant from each other compared to the second example but the changes along the path are smooth. Moreover, in the second example, X_2 and X_3 are very close and we observe the $\alpha(t)$ between them is quasi-constant. Same between X_3 and X_4, red and blue points are quasi-fixed while the green points are moving in the right direction smoothly. Another way to visualize the smoothness of α in the second example is to plot the norm of the first and second derivatives. We illustrate this idea in Fig. 5.7(top). In the same figure, we show the norm of the first and second derivatives of a piecewise geodesic path for comparison. We can easily check that the proposed solution is better.

5.4 Interpolation Problem on the Stiefel Manifold

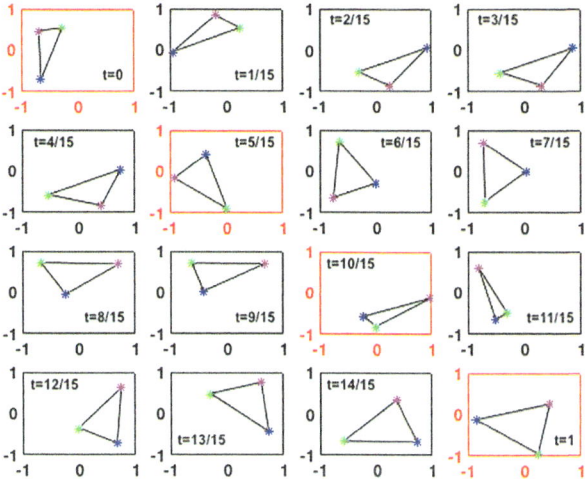

Fig. 5.5 Example of $\alpha(t), t \in \{0, \frac{1}{15}, \ldots, 1\}$ on $\mathcal{M} = St(3, 2)$. The original observations are given on the diagonal (red) at $t \in \{0, \frac{1}{3}, \frac{2}{3}, 1\}$. All $\alpha(t)$ are uniformly spaced in $[0, 1]$

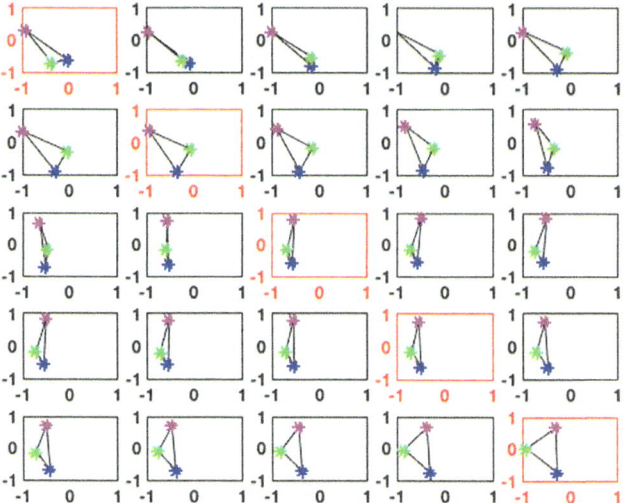

Fig. 5.6 Example of $\alpha(t), t \in \{0, \frac{1}{25}, \ldots, 1\}$ on $\mathcal{M} = St(3, 2)$. The original observations are given on the diagonal (red) and all $\alpha(t)$ are uniformly spaced in $[0, 1]$

5.4.3.3 Case 3: Application on Real Data

A problem of great importance, and of current interest in the scientific community, is studying and exploiting electrocardiograms as a non-invasive methodology to detect heart diseases. While there are many successful works in the literature to study cardiological problems by means of static observations, there is still a need for

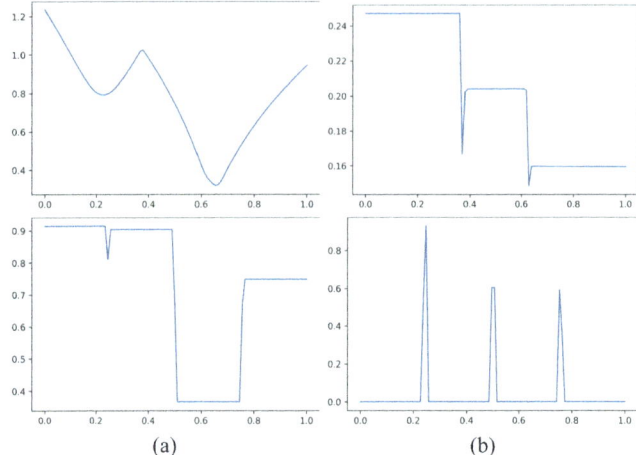

Fig. 5.7 The norm of the first derivative (**a**) and the second derivative (**b**) of interpolating paths $\alpha(t)$, $t \in [0, 1]$ using the same example in Fig. 5.6. Top is the proposed solution and bottom is a piecewise geodesic path

mathematical models to study dynamic spatio-temporal changes. Thus, evolution between subjects or temporal observations for the same subject can help capture cross-sectional and functional variabilities [22]. In a medical context, such models can assist physicians in the interpretation of sequences. For example, they can be used to highlight changes during different temporal observations of the same subject. In this spirit, we present a new trajectory-based approach that can smoothly interpolate sequences observed from different subjects to study inter-subject variability (stages). We remind that the main goal is restricted to illustrating potential real applications without any medical interpretation of results.

We consider a real dataset of vectocardiograms (VCG) from children with ages varying between 2 and 19 where each observation $X_i = (x_i^1, x_i^2)$ belongs to $\mathcal{M} = St(3, 2)$ [23]. Each element describes the unitary vectors: x_i^1 as the direction to the apex and x_i^2 as the direction of motion. Figure 5.8 represents each matrix observation $X_i = (x_i^1, x_i^2)$ by a points in $\mathcal{M} = St(3, 2)$. They are then interpolated by a path α in $\mathcal{M} = St(3, 2)$. The observations are displayed in red on the diagonal and all points $\alpha(t)$ are uniformly spaced in $[0, 1]$. To better visualize the dynamics along α, we show the norm of the first and second derivatives in Fig. 5.9 (top). We also show a piecewise geodesic path for comparison (bottom).

5.4 Interpolation Problem on the Stiefel Manifold

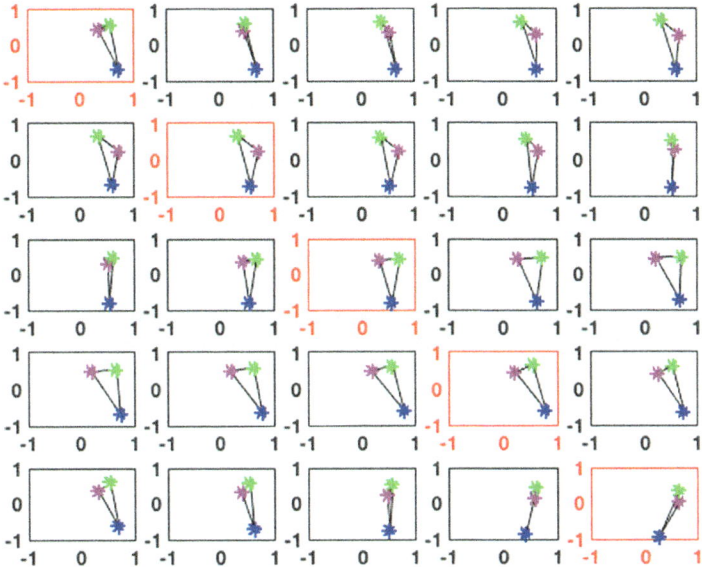

Fig. 5.8 Example of $\alpha(t), t \in \{0, \frac{1}{25}, \ldots, 1\}$ on $\mathcal{M} = St(3, 2)$. The original observations are given on the diagonal (red) and all $\alpha(t)$ are uniformly spaced in $[0, 1]$

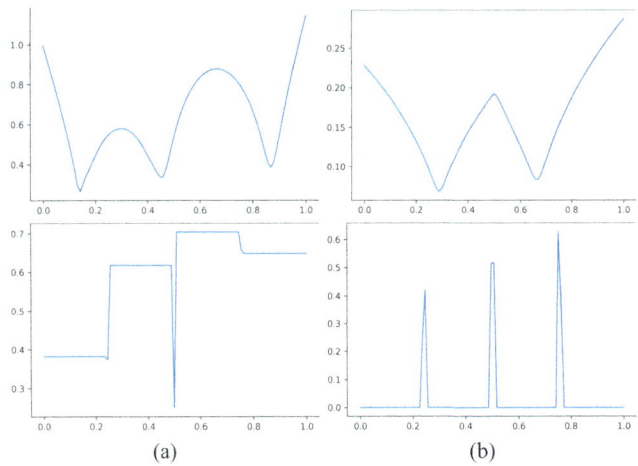

Fig. 5.9 The norm of the first derivative (**a**) and the second derivative (**b**) of interpolating paths $\alpha(t), t \in [0, 1]$ using the same example in Fig. 5.8. Top is the proposed solution and bottom is piecewise geodesic path

5.5 Interpolation Problem on the Grassmann Manifold

In this section, we demonstrate that in the case of the Grassmann Manifold $\mathcal{M}$, the Bézier spline α exhibits a C^2 smoothness, a consequence of the manifold's elegant properties. To elaborate, let $(N+1)$ distinct data points $X_0, ..., X_N$ in $\mathcal{M}$ be associated with time instants $t_i = i, i = 0, ..., N$. The objective is to construct a Bézier spline $\alpha : [0, N] \to \mathcal{M}$ that interpolates the data points X_i for $i = 0, ..., N$ while ensuring a C^2 differentiability condition at the knot points. Similar to the previous section, the Bézier spline α is recursively constructed through a chain of Bézier curves α_j, $j \in \{2, 3\}$ defined by a set of $(N-1)$ control points $(\widehat{Y}_i^-, \widehat{Y}_i^+)$ on the left and right sides of the interpolation point X_i for $i = 1, ..., (N-1)$. The proposed algorithm for determining the optimal control points of the C^2 Bézier spline in $\mathcal{M}$ comprises two distinct phases:

1. **Phase** 1: We construct a C^1 interpolation Bézier spline, denoted as $\alpha : [0, N] \to \mathcal{M}$, within $\mathcal{M}$. To achieve this, for each $i = 1, ..., N-1$, we utilize the Riemannian logarithmic map, given by Eq. 5.10, to transfer the data $X_0, ..., X_N$ to each tangent space $T_{X_i}\mathcal{M}$. The mapped data is then represented as $S^i = (S_0^i, ..., S_N^i)$, where $S_k^i = \text{Log}_{X_i}(X_k)$ for $k = 0, ..., N$, and $i = 1, ..., (N-1)$. Subsequently, employing the equations governing the control points of the C^2 Bézier spline on $\mathbb{R}^n$ provided in Algorithm 4, we construct the C^2 Bézier curve β_i in each tangent space $T_{X_i}\mathcal{M}$ for $i = 1, ..., N-1$. Finally, using the Riemannian exponential map Exp_{X_i} defined on $\mathcal{M}$ by Eq. 5.9, we transport the control points of the C^2 Bézier curve β_i from the tangent space $T_{X_i}\mathcal{M}$ to the matrix manifold $\mathcal{M}$, providing the control points for the desired C^1 interpolation Bézier spline α within $\mathcal{M}$. The summarized result is presented in Theorem 5.2.

Theorem 5.2 *Given $X_0, ..., X_N$ a set of $(N+1)$ points in $\mathcal{M}$, and $B^i = [(B_1^i)^-, ..., (B_{N-1}^i)^-]^T$ a matrix of size $(n(N-1) \times n)$ such that each row B_i^i of B^i contains optimal control points of the C^2 Bézier spline β_i in each tangent space $T_{X_i}\mathcal{M}$, for $i = 1, .., N-1$. Then, the Bézier spline $\alpha : [0, N] \to \mathcal{M}$ interpolating the data X_i is of class C^1 and is uniquely defined by the set of control points $\widehat{Y} = [\widehat{Y}_1^-, ..., \widehat{Y}_{N-1}^-]^T \in \mathbb{R}^{n(N-1) \times n}$ given by:*

$$\widehat{Y}_i^- = \text{Exp}_{X_i}(B_i^i), \; i = 1, ..., N-1, \tag{5.17}$$

Proof The proof is similar to the one of Theorem 3.2. □

Remark 5.3 The interpolation point X_N is modified under the C^2 differentiability condition of the curve β_i on $T_{X_i}\mathcal{M}$, for $i = 1, ..., N-1$, therefore the point X_N is changed and the new $(N+1)$ interpolation points on $\mathcal{M}$ is given by:

$$\widetilde{X}_k = \text{Exp}_{X_i}(\widetilde{S}_k^i), \; k = 0, ..., N; \; i = 1, ..., N-1. \tag{5.18}$$

where $\widetilde{S} = [\widetilde{S}_0^i, ..., \widetilde{S}_N^i]^T$ a matrix of size $n(N+1) \times n$ containing the new $(N+1)$ interpolation points in each tangent space $T_{X_i}\mathcal{M}$.

It is important to notice that at this first step, the Bézier spline α is only of class C^1. The next goal is to show how to arrange α to satisfy the C^2 differentiability condition at points $X_i, i = 1, ..., N$.

2. **Phase** 2: At this stage, we capitalize on the advantageous properties of symmetric spaces. A detailed examination of the relationship between global symmetries at interpolation points and the C^2 differentiability of the Bézier spline on symmetric spaces mirrors the analysis provided in Lemma 4.2 and Theorem 4.1 in the previous chapter. These analyses involve the computation of the derivative of the geodesic symmetry, as given by Eq. 5.14), and the derivative of the Riemannian exponential map $\mathrm{Exp} X i$ defined in $\mathcal{M}$ by Eq. 5.9. Let Y_i^- and Y_i^+ denote the new control points on the left and right sides of the interpolation point $\tilde{X}_i$, defining the Bézier spline α in $\mathcal{M}$. The fundamental approach to determining the control points Y_i^-, for $i = 1, ..., N - 1$, closely mirrors the procedure in the Euclidean case $\mathbb{R}^n$. Specifically, we may know Y_1^- (and consequently Y_1^+ by the C^1 differentiability condition in $\mathcal{M}$) and aim to iteratively define Y_i^- for $i = 2, ..., N - 1$ (along with Y_i^+, similar to the determination of Y_1^+).

Theorem 5.3 *Under the hypotheses of Theorem 5.2, the Bézier spline $\alpha : [0, N] \to \mathcal{M}$ is C^2 and is uniquely defined by the set of control points given by the row of the matrix $Y = [Y_1^-, ..., Y_{N-1}^-]^T \in \mathbb{R}^{n(N-1) \times n}$ by*

1. $Y_1^- = \mathrm{Exp}_{\tilde{X}_1}((B_1^1)^-)$,
2. $Y_2^- = \mathrm{Exp}_{Y_1^+}\left(\frac{1}{3}\left((d\varphi_{\tilde{X}_1})_{Y_1^-}\left(\dot{\gamma}(1, \tilde{X}_0, Y_1^-)\right) - 4\dot{\gamma}(0, Y_1^-, \tilde{X}_1)\right)\right)$,
3. $Y_{i+1}^- = \mathrm{Exp}_{Y_i^+}\left(\left((d\varphi_{\tilde{X}_i})_{Y_i^-}\left(\dot{\gamma}(1, Y_{i-1}^+, Y_i^-)\right) - 2\dot{\gamma}(0, Y_i^-, \tilde{X}_i)\right)\right)$.
$i = 2, ..., N - 2$.

where $d\varphi_{\tilde{X}_i}$ represent the derivative of the symmetry map $\varphi_{\tilde{X}_i}$ at a point Y_i^-, $i = 1, ..., N - 1$.

Proof The argument follows a structure similar to the proof of Theorem 3.3. To facilitate comprehension, we recapitulate the key points:

1. We calculate the covariant derivative of a tangent vector along a curve and establish a clear C^2 condition on symmetric spaces, expressed in terms of the derivative of the exponential and symmetry functions. This outcome, coupled with Theorem 5.2, streamlines the differentiation of the inverse of the exponential map.
2. The derivative of the symmetry is then represented as a function of the tangent vector along the Bézier spline α_i^j at $t = 0$ and $t = 1$. This simplifies the C^2 condition outlined in Theorem 4.1, aiding us in deriving an explicit expression for the control points responsible for generating the Bézier spline α.

□

We outline the various steps of the proof in Algorithm 9.

Algorithm 9 C^2 interpolating Bézier spline on the Grassmann manifold $\mathcal{M}$.

Input: $N \geq 3$, $\tilde{X} = [\tilde{X}_0, ..., \tilde{X}_N]^T$ a matrix of size $n(N+1) \times n$ containing the $(N+1)$ interpolation points in $\mathcal{M}$.
Output: Y.
1: Calculate $\widehat{Y} = [\widehat{Y}_1^-, ..., \widehat{Y}_{N-1}^-]^T$ a matrix of size $n(N-1) \times n$ containing the $(N-1)$ control points of the C^2 Bézier curve β_i on $T_{X_i}\mathcal{M}$ using Algorithm 4.
2: Set $Y_1^- = \widehat{Y}_1^-$.
3: Calculate control point Y_1^+:
4: $Y_1^+ = \mathrm{Exp}_{\tilde{X}_1}(-\frac{2}{3}\mathrm{Exp}_{\tilde{X}_1}^{-1}(Y_1^-))$
5: Calculate control point Y_2^-:
6: $Y_2^- = \mathrm{Exp}_{Y_1^+}\left(\frac{1}{3}\left((d\varphi_{\tilde{X}_1})_{Y_1^-}\left(\dot{\gamma}(1, \tilde{X}_0, Y_1^-)\right) - 4\dot{\gamma}(0, Y_1^-, \tilde{Y}_1)\right)\right)$
7: **for** $i = 2 : N - 2$ **do**
8:     $Y_i^+ = \mathrm{Exp}_{\tilde{X}_i}(-\mathrm{Exp}_{\tilde{X}_i}^{-1}(Y_i^-))$
9:     $Y_{i+1}^- = \mathrm{Exp}_{Y_i^+}\left(\left((d\varphi_{\tilde{X}_i})_{Y_i^-}\left(\dot{\gamma}(1, Y_{i-1}^+, Y_i^-)\right) - 2\dot{\gamma}(0, Y_i^-, \tilde{X}_i)\right)\right)$
10: **end for**
11: Calculate control point Y_{N-1}^+:
12: $Y_{N-1}^+ = \mathrm{Exp}_{\tilde{X}_{N-1}}(-\frac{2}{3}\mathrm{Exp}_{\tilde{X}_{N-1}}^{-1}(Y_{N-1}^-))$
13: **return** Y,

References

1. Man Lui,Y.: Advances in matrix manifolds for computer vision. ImageVis. Comput. 30(6-7), 380-388, 2012.
2. Zhang, D. and Balzano, L.:Global convergence of a Grassmannian gradient descent algorithm for subspace estimation. Proceedings of the 19th international conference on artificial intelligence and statistics, AISTATS, pp. 1460-1468. Cadiz, Spain, 2016.
3. Nguyen, T.S.: A real time procedure for affinely dependent parametric model order reduction using interpolation on Grassmann manifolds. Int. J. Numer. Meth. Eng. 93(8), 818-833, 2013.
4. Absil, P.A and Mahony, R. and Sepulchre,R. : Optimization Algorithms on Matrix Manifolds, Princeton University Press, 2008.
5. Chakraborty, R. and Vemuri, B. C.: Statistics on the compact Stiefel manifold: Theory and Applications, The Annals of Statistics, 47, (1), 2017.
6. Clément, P. and Guillaume, B. and Vincent,C. : Improved Time-Series Clustering with UMAP dimension reduction method, 25th International Conference on Pattern Recognition (ICPR), 5658-5665, 2021.
7. David, S. L and Sahib, A. and Narr, K. and Nunez, E. and Joshi, S. H.: Global Diffeomorphic Phase Alignment of Time-Series from Resting-State fMRI Data, Medical Image Computing and Computer Assisted Intervention - MICCAI, Lecture Notes in Computer Science, Springer, Cham ,12267, 518–527, 2020.
8. Edelman, A. and Arias, T. A. and Smith,S. T.: The geometry of Algorithms with Orthogonality Constraints, SIAM J. Matrix Anal. Appl, 20 (2), 303-353, 1998.
9. Batzies,E. and Hüper, K. and Machado, L. and Silva Leite,F.: Geometric mean and geodesic regression on Grassmannians, Linear Algebra and its Applications, 83-101, 466, 2015.
10. Hüper, K. and Helmke, U. and Herzberg,S.: On the computation of means on Grassmann manifolds, Pro-ceedings of the 19th International Symposium on Mathematical Theory of Networks and Systems, MTNS, 2439-2441, 2010.

References

11. Nishimori, Y. and Akaho,S. : Learning algorithms utilizing quasi-geodesic flows on the Stiefel manifold, Neurocomputing, 67, 106 - 135, 2005.
12. Rentmeesters, Q.: A gradient method for geodesic data fitting on some symmetric Riemannian manifolds, IEEE CDC-ECC, 7141-7146, 2011.
13. Zimmermann, R.: A matrix-algebraic algorithm for the Riemannian logarithm on the Stiefel manifold under the Canonical metric, SIAM J. Matrix Anal. Appl, 38(2), 322-342, 2017.
14. Bryner, D.: Endpoint Geodesics on the Stiefel Manifold Embedded in Euclidean Space, SIAM J. Matrix Anal. Appl, 38, 1139-1159, 2017.
15. Sundaramoorthi, G. and Mennucci, A. and Soatto, S. and Yezzi, A.J: A New Geometric Metric in the Space of Curves, and Applications to Tracking Deforming Objects by Prediction and Filtering, SIAM JIS, 4(1), 109-145, 2011.
16. Hüper, K. and Silva Leite,F. : On the geometry of rolling and interpolation curves on S^n, $SO(n)$, and Grassmann manifolds, J. Dyn. Control Syst, 13(4), 467-502, 2007.
17. Krakowski, K.A. and Machado, L. and Silva Leite, F. and Batista, J.: A modified Casteljau algorithm to solve interpolation problems on Stiefel manifolds, J. Comput. Appl. Math, 311, 84-99, 2017.
18. Hong, Y. and Kwitt, R. and Singh, N. and Davis, B. and Vasconcelos, N. and Niethammer, M.: Geodesic Regression on the Grassmannian, European Conference on Computer Vision ECCV, 632-646, 2014.
19. Dryden, I. L. and Mardia, K. V.: Statistical shape analysis with applications in R, John Wiley and Sons, Chichester, 2016.
20. Kyle, G. and Anuj, S. and Xiuwen, L. and Paul, V.D.: Efficient algorithms for inferences on Grassmann manifolds, Proceedings of 12 th IEEE Workshop on Statistical Signal Processing, 315-318, 2003.
21. Bendory, T. and Dekel, S. and Feuer, A.: Super-resolution on the Sphere using Convex Optimization, arXiv:1412.3282, 2014.
22. Cholaquidis, A. and Fraiman, R. and Moreno, L.: Weighted lens depth: Some applications to supervised classification, arXiv preprint arXiv:2011.11140, 2020.
23. Pal, S. and Sengupta, S. and Mitra, R. and Banerjee, A.:A bayesian approach for analyzing data on the stiefel manifold, arXiv:1907.04303, 2019.

Chapter 6
Spline Interpolation on the Manifold of Probability Measures

In this chapter, we present an efficient and accurate algorithm for constructing a C^2 Bézier spline that interpolates a given ordered set of data points on the space of probability measures $\mathcal{P}_+$, equipped with the Fisher–Rao metric. The distinctive aspect of the proposed method lies in its exploration of the inherent Riemannian structure within the space of probability measures, making the solution computationally feasible.

The initial step involves deriving explicit theoretical expressions for crucial geometric structures on $\mathcal{P}_+$, including the Levi-Civita connection, minimal geodesics, parallel transport, exponential, and logarithm maps. Subsequently, by combining adequate standard interpolation methods and newly geodesic operations, an optimal interpolating Bezier spline, in the sense of an appropriate energy functional is constructed on $\mathcal{P}_+$. Finally, we showcase the method's high accuracy and efficiency through numerous numerical examples.

6.1 Problem Formulation

Consider $\mu^{(0)}, \ldots, \mu^{(N)}$ as an indexed set of probability measures associated to a set of observation time $t_0, \ldots, t_N$, for $i = 0, \ldots, N$. The purpose is to estimate a path $\gamma : [0, N] \to \mathcal{P}_+$ that minimizes the following functional

$$E(\gamma) = \frac{\lambda}{2} \int_{t_0}^{t_N} <\frac{D^2\gamma(t)}{Dt^2}, \frac{D^2\gamma(t)}{Dt^2}>_F + \frac{1}{2}\sum_{i=0}^{N} d_F^2(\gamma(t_i), \mu^{(i)}), \qquad (6.1)$$

where $(\lambda > 0)$ is a smoothing parameter, $<.,.>_F$ and d_F denote the Fisher–Rao metric and the geodesic distance on $\mathcal{P}_+$. In this chapter, we aim to find γ, a solution of the optimization problem (6.1) on $\mathcal{P}_+$, that satisfies the following properties: (i) $\gamma(t_i) = \mu^{(i)}$, (ii) γ is of class C^2.

Recently, significant attention has been directed toward the space of probability measures $\mathcal{P}_+$, due to its interesting geometric structure highlighted by the field of information geometry in the seminal works of Amari and Nagaoka [1]. In essence, information geometry employs the tools of differential geometry to explore spaces of probability measures by treating them as manifolds referred to as statistical manifolds. Many different metrics and semi-metrics on $\mathcal{P}_+$ have been studied, including $\mathbb{L}^1$, log-Euclidean, $\mathbb{L}^2$, Jensen–Shannon, and Wasserstein metrics [2–5]. This space has been extensively studied in the context of linear regression [6, 7], transport of statistical models [8], estimation [9], and barycenters [10, 11], leading to computational advances in statistical analysis. Moreover, the interpolation problem on the space of probability measures $\mathcal{P}_+$ equipped with Wasserstein metrics has been concurrently and independently considered by various authors, as seen in [12–14].

As far as we know, this framework is the first to address the interpolation problem on the space of probability measures $\mathcal{P}_+$ within the context of splines. In this chapter, we focus on the finite-dimensional space $\mathcal{P}_+$ with dimension n. Therefore, this space can be represented as the simplex in the the Euclidean space $\mathbb{R}^{n+1}$. Additionally, we equip $\mathcal{P}_+$ with the Fisher metric, a Riemannian metric described in matrix form by the well-known Fisher information matrix introduced in [15]. It's worth noting that obtaining closed-form expressions for the Fisher–Rao distance can be a challenging task, as it requires solving a Riemannian geodesic equation with boundary conditions. Having that in mind, in this paper, we provide a closed-form expression for the Fisher–Rao distance to evaluate dissimilarities between probability measures. Furthermore, we establish a connection between the Fisher geometry and the geometry of the finite-dimensional sphere. Moreover, we derive explicit expression for the Christoffel symbols and, consequently, the Levi-Civita connection associated with the Fisher–Rao metric, enabling the computation of geodesic curves joining two points on $\mathcal{P}_+$. Consequently, we present an exact equation for the Levi-Civita parallel transport of a tangent vector along a geodesic curve joining two probability measures on $\mathcal{P}_+$. Furthermore, we establish that $\mathcal{P}_+$ is a locally symmetric space. Leveraging these favorable results, We extend the methodology presented in Chap. 2 to generate an algorithm for constructing a measure-interpolating spline of class C^2 on $\mathcal{P}_+$. We demonstrate the effectiveness of the proposed approach with potential applications.

6.2 Geometry of the Space of Probability Measures

In this section, we develop Riemannian calculus on $\mathcal{P}_+$ with the Fisher–Rao metric, deriving various geometric concepts, such as geodesics, exponential maps, logarithm maps, and the Levi-Civita parallel transport.

6.2.1 Manifold Structure

Let $I = \{1, \ldots, n, n+1\}$, $n \in \mathbb{N}$, be a finite sample space. Let $\mathcal{F}(I) = \{f : I \to \mathbb{R}\}$ be the algebra of real functions on I. Its unity function $\mathbb{1}_I$ or simply $\mathbb{1}$ is given by $\mathbb{1}(i) = 1$, for $i = 1, \ldots, n, n+1$. A canonical basis of $\mathcal{F}(I)$ is defined by

$$e_i(j) = \begin{cases} 1, & \text{if } i = j, \\ 0, & \text{otherwise,} \end{cases} \tag{6.2}$$

and hence, every $f \in \mathcal{F}(I)$ has the representation

$$f = \sum_{i \in I} f^i e_i, \tag{6.3}$$

where $f^i = f(i)$. We will denote by $\mathcal{S}(I)$ the dual space of $\mathcal{F}(I)$, the space of $\mathbb{R}$-valued linear forms on $\mathcal{F}(I)$. With the Riesz representation theorem, this vector space is interpreted as the vector space of signed measures on I, namely

$$\mathcal{S}(I) = \{\mu : \mathcal{F}(I) \to \mathbb{R} \mid \mu = \sum_{i \in I} \mu_i \delta^i\}, \tag{6.4}$$

where $\mu_i = \mu(e_i)$ and δ^i is considered as the Dirac measure supported at $i \in I$. It is also shown that $\mathcal{S}(I)$ is a smooth manifold. Besides we have a vector space isomorphism between the space $\mathcal{F}(I)$ and $\mathcal{S}(I)$, given by

$$\begin{aligned} \mathcal{F}(I) &\longrightarrow \mathcal{S}(I) \\ f &\longmapsto f\mu := \sum_{i \in I} f^i \mu_i \delta^i. \end{aligned} \tag{6.5}$$

The inverse is the Radon–Nikodym derivative with respect to μ, denoted as ϕ_μ,

$$\begin{aligned} \phi_\mu : \mathcal{S}(I) &\longrightarrow \mathcal{F}(I) \\ \nu = \sum_{i \in I} \nu_i \delta^i &\longmapsto \frac{d\nu}{d\mu} := \sum_{i \in I} \frac{\nu_i}{\mu_i} e_i. \end{aligned} \tag{6.6}$$

In particular, the tangent space at the point $\mu \in \mathcal{S}(I)$ is given by

$$T_\mu \mathcal{S}(I) = \{\mu\} \times \mathcal{S}(I). \tag{6.7}$$

Let us consider the following submanifolds of $\mathcal{S}(I)$:

$$\mathcal{S}_\epsilon(I) = \left\{ \mu = \sum_{i \in I} \mu_i \delta^i \mid \sum_{i \in I} \mu_i = \epsilon, \ \epsilon \in \mathbb{R} \right\},$$

and

$$\mathcal{M}_+(I) = \{ \mu \in \mathcal{S}(I) \mid \mu_i > 0, \ \forall \ i \in I \},$$

the space of finite strictly positive measures on I.

Definition 6.1 A probability measure on a finite sample space I is a map $\mu : I \to \mathbb{R}$ defined for any $A \subset I$ by $\mu(A) = \sum_{i \in A} \mu_i$ and which satisfies:

1. For all $i \in I$, $\mu_i \geq 0$ and $\mu(\emptyset) = 0$.
2. $\sum_{i \in I} \mu_i = 1$.
3. $\mu(\{i\}) = \mu_i$.

We denote by $\mathcal{P}_+(I)$ the space of strictly positive probability measures on I,

$$\mathcal{P}_+(I) = \left\{ \mu = \sum_{i \in I} \mu_i \delta^i \mid \mu_i > 0, \ \forall i \in I, \text{ and } \sum_{i \in I} \mu_i = 1 \right\}.$$

We check at once that $\mathcal{P}_+(I) \subset \mathcal{M}_+(I) \subset \mathcal{S}(I)$. Therefore, as an open submanifold of $\mathcal{S}(I)$, $\mathcal{M}_+(I)$ has the same tangent space at the point $\mu \in \mathcal{M}_+(I)$. $\mathcal{P}_+(I)$ is a submanifold of $\mathcal{S}(I)$, and clearly, for $\mu \in \mathcal{P}_+(I)$, we have:

$$T_\mu \mathcal{P}_+(I) = \{\mu\} \times \mathcal{S}_0(I)$$
$$= \{(\mu, v) \mid \mu \in \mathcal{P}_+(I) \text{ and } v \in \mathcal{S}_0(I)\}.$$

We want to endow $\mathcal{P}_+(I)$ with a Riemannian metric. To this end, we define a local coordinate map on $\mathcal{P}_+(I)$. Let U be an open set of $\mathbb{R}^n$ given by

$$U = \left\{ x = (x_1, \ldots, x_n) \in \mathbb{R}^n \mid x_i > 0, \forall i \in I, \text{ and } \sum_{i=1}^n x_i < 1 \right\}.$$

We define a map φ as

$$\varphi : \mathcal{P}_+(I) \longrightarrow U,$$
$$\mu = \sum_{i \in I} \mu_i \delta^i \longmapsto (\varphi^1(\mu), \ldots, \varphi^n(\mu)) = (x^1(\mu), \ldots, x^n(\mu)),$$

such that $(\varphi^1(\mu), \ldots, \varphi^n(\mu)) = (\mu_1, \ldots, \mu_n)$. Clearly, φ is an homomorphism and its inverse is given by

6.2 Geometry of the Space of Probability Measures

$$\varphi^{-1} : U \longrightarrow \mathcal{P}_+(I),$$

$$(x_1, \ldots, x_n) \longmapsto \mu = \sum_{i=1}^n x_i \delta^i + \left(1 - \sum_{i=1}^n x_i\right) \delta^{n+1}.$$

Given a point $\mu \in \mathcal{P}_+(I)$, let $\left.\dfrac{\partial}{\partial x^i}\right|_\mu$ be the tangent vector at μ given by

$$\left.\frac{\partial}{\partial x^i}\right|_\mu = \left.\frac{\partial}{\partial x_i}\right|_{\varphi(\mu)} \varphi^{-1} = \left(\delta^i - \delta^{n+1}\right), \quad \text{for } i = 1, \ldots, n.$$

Thus, $\left\{\left.\dfrac{\partial}{\partial x^i}\right|_\mu, i = 1, \ldots, n\right\}$ define a local frame field of $T_\mu \mathcal{P}_+(I)$ at a point $\mu \in \mathcal{P}_+(I)$. Similarly we can define the dual basis of $\left.\dfrac{\partial}{\partial x^i}\right|_\mu$, the basis of the cotangent bundle $T_\mu^* \mathcal{P}_+(I) = \{\mu\} \times (\mathcal{F}(I)/\mathbb{R})$ by $dx^i = e_i + \mathbb{R}, \quad i = 1, \ldots, n$.

Remark 6.1 Let $\mu \in \mathcal{P}_+(I)$ and $v = \sum_{i \in I} v_i \delta^i \in T_\mu \mathcal{P}_+(I)$. It can be easily seen that, for $v \in \mathcal{S}_0(I)$

$$v = \sum_{i=1}^{n+1} v_i \delta^i = \sum_{i=1}^n v_i \delta^i - \sum_{i=1}^n v_i \delta^{n+1}$$

$$= \sum_{i=1}^n v_i (\delta^i - \delta^{n+1}) = \sum_{i=1}^n v_i \frac{\partial}{\partial x^i}.$$

$S(I)$ is a finite-dimensional linear space, and therefore, it can be naturally equipped with a metric. For $v, w \in T_\mu S(I)$, we define the inner product as

$$<v, w>_\mu = \mu \left(\frac{dv}{d\mu} \cdot \frac{dw}{d\mu}\right) = \sum_i \frac{v_i w_i}{\mu_i}, \tag{6.8}$$

where $\dfrac{dv}{d\mu} = \sum_{i \in I} \dfrac{v_i}{\mu_i} e_i \in \mathcal{F}(I)$, represents a simple version of the Radon–Nikodym derivative with respect to μ. This metric induces a metric on $\mathcal{M}_+(I)$. The probability manifold $\mathcal{P}_+(I)$ as a submanifold of $\mathcal{M}_+(I)$, is endowed with the Fisher–Rao metric. Hence, following the geometry structures in $\mathcal{M}_+(I)$ equipped with Fisher information metric, we derive the corresponding ones in $\mathcal{P}_+(I)$.

Definition 6.2 Let μ be a probability measure in $\mathcal{P}_+(I)$. Given two tangents vectors v and w in $T_\mu \mathcal{P}_+(I)$, the Fisher–Rao metric $\mathfrak{g}_\mu : T_\mu \mathcal{P}_+(I) \times T_\mu \mathcal{P}_+(I) \to \mathbb{R}$ is defined by

$$\mathfrak{g}_\mu(v, w) = \sum_{i \in I} \frac{v_i w_i}{\mu_i}, \tag{6.9}$$

and $||v||_\mu = \sqrt{\mathfrak{g}_\mu(v,v)}$. With respect to the coordinate map $(\mathcal{P}_+(I), \varphi)$, the Fisher–Rao metric is expressed as

$$g_{ij}(\mu) = \begin{cases} \frac{1}{\mu_i} + \frac{1}{\mu_{i+1}}, & \text{if } i = j, \\ \frac{1}{\mu_{n+1}}, & \text{otherwise,} \end{cases}$$

for $i, j = 1, \ldots, n$. And the components of the inverse matrix are given by

$$g^{ij}(\mu) = \begin{cases} \mu_i(1 - \mu_i), & \text{if } i = j, \\ -\mu_i \mu_j, & \text{otherwise.} \end{cases}$$

6.2.2 Fisher–Rao Metric on $\mathcal{P}_+$

Let $\mathcal{X}(\mathcal{P}_+(I))$ denote the set of smooth vector fields on $\mathcal{P}_+(I)$. Essentially, at each point $\mu \in \mathcal{P}_+(I)$, the Levi-Civita connection associated with the Fisher–Rao metric $\nabla : \mathcal{X}(\mathcal{P}_+(I)) \times \mathcal{X}(\mathcal{P}_+(I)) \to \mathcal{X}(\mathcal{P}_+(I))$ gives a new vector field, notated $\nabla_X Y$, telling us how the vector field Y is changing in the direction X and satisfying for all $X, Y, Z \in \mathcal{X}(\mathcal{P}_+(I))$,

$$\begin{cases} X\mathfrak{g}(Y, Z) = \mathfrak{g}(\nabla_X Y, Z) + \mathfrak{g}(Y, \nabla_X Z), \\ \nabla_X Y - \nabla_Y X = [X, Y]. \end{cases} \quad (6.10)$$

In the local coordinate map $(\mathcal{P}_+(I), \varphi)$, the Levi-Civita connection is defined by the Christoffel symbols $\Gamma_{ij}^k : \mathcal{P}_+(I) \to \mathbb{R}$ such that $\nabla_{\partial x_i} \partial x_j = \Gamma_{ij}^k \partial x_k$.

Proposition 6.1 *With respect to the local coordinate map $(\mathcal{P}_+(I), \varphi)$, the Christoffel symbols associated with the Fisher–Rao metric are given by*

$$\Gamma_{ij}^k = \begin{cases} \dfrac{1}{2} \dfrac{x_k}{1 - \sum_{h=1}^n x_h}, & i \neq j, \\ \dfrac{1}{2} \dfrac{x_k}{1 - \sum_{h=1}^n x_h} + \dfrac{1}{2} \dfrac{x_k}{x_i}, & i = j \neq k, \\ \dfrac{1}{2} \dfrac{x_k}{1 - \sum_{h=1}^n x_h} - \dfrac{1}{2} \dfrac{1 - x_k}{x_k}, & i = j = k, \end{cases} \quad (6.11)$$

Proof The smooth functions Γ_{ij}^k are easily computed through the characterization of the Levi-Civita connection by the Koszul formula obtained from (6.10) computed for all the circular permutations of $X, Y, Z \in \mathcal{X}(\mathcal{P}_+(I))$,

$$\begin{aligned} \mathfrak{g}(\nabla_X Y, Z) =& \frac{1}{2} \{ X\mathfrak{g}(Y, Z) + Y\mathfrak{g}(Z, X) - Z\mathfrak{g}(X, Y) \\ & + \mathfrak{g}([X, Y], Z) - \mathfrak{g}([Y, Z], X) - \mathfrak{g}([X, Z], Y) \}. \end{aligned} \quad (6.12)$$

6.2 Geometry of the Space of Probability Measures

Now, in the Koszul formula we set $X = \partial x_i$, $Y = \partial x_j$ and $Z = \partial x_l$. We get

$$\Gamma_{ij}^k = \frac{1}{2} \sum_{l=1}^n g^{kl} \left(g_{il,j} + g_{jl,i} - g_{ij,l} \right), \tag{6.13}$$

for $i, j, k \in \{1, \ldots, n\}$, where $g_{il,j} = \dfrac{\partial g_{il}}{\partial x_j}$, $g_{jl,i} = \dfrac{\partial g_{jl}}{\partial x_i}$, and $g_{ij,l} = \dfrac{\partial g_{ij}}{\partial x_l}$. In the local coordinate system, the Fisher–Rao metric and its inverse are given by

$$g_{ij} = \begin{cases} \dfrac{1}{x_i} + \dfrac{1}{1 - \sum_{h=1}^n x_h}, & \text{if } i = j, \\ \dfrac{1}{1 - \sum_{h=1}^n x_h}, & \text{if } i \neq j, \end{cases} \tag{6.14}$$

$$g^{ij} = \begin{cases} x_i(1 - x_i), & \text{if } i = j, \\ -x_i x_j, & \text{if } i \neq j, \end{cases} \tag{6.15}$$

for $i, j = 1, \ldots, n$. Now if we take the derivative of Eq. 6.14 by x_l, we get

$$g_{ij,l} = \begin{cases} -\dfrac{1}{(x_i)^2} + \dfrac{1}{(1 - \sum_{h=1}^n x_h)^2}, & \text{if } i = j = l, \\ \dfrac{1}{(1 - \sum_{h=1}^n x_h)^2}, & \text{otherwise.} \end{cases} \tag{6.16}$$

Replace Eq. 6.16 in Eq. 6.13, the formula follows. $\square$

Definition 6.3 Let $X \in \mathcal{X}(\mathcal{P}_+(I))$ be a vector field on $\mathcal{P}_+(I)$. Then in the local coordinate $(\mathcal{P}_+(I), \varphi)$, we have the representation $X = \sum_{i=1}^n X_i \partial x_i$. X is called a constant vector field on $\mathcal{P}_+(I)$ if all X_i are independent of μ.

Theorem 6.1 *Given two constant vector fields X, Y on $\mathcal{P}_+(I)$, the Levi-Civita connection at $\mu \in \mathcal{P}_+(I)$ is given by*

$$\nabla_X Y(\mu) = -\frac{1}{2} \left(\frac{dX}{d\mu} \frac{dY}{d\mu} - \mathfrak{g}_\mu(X, Y) \right) \mu. \tag{6.17}$$

Proof Let $X = \sum_{i \in I} X_i \delta^i$, $Y = \sum_{i \in I} Y_i \delta^i$ and $Z = \sum_{i \in I} Z_i \delta^i$ be constant vector fields on $\mathcal{P}_+(I)$. Thus, we get $[X, Y] = [Y, Z] = [X, Z] = 0$ and consequently Eq. 6.12 gives

$$\mathfrak{g}(\nabla_X Y, Z) = \frac{1}{2} \{ X \mathfrak{g}(Y, Z) + Y. \mathfrak{g}(X, Z) - Z. \mathfrak{g}(X, Y) \}. \tag{6.18}$$

Set $\mu = \sum_{i \in I} \mu_i \delta^i \in \mathcal{P}_+(I)$, and $\alpha(t) = \mu + vt$, a curve on $\mathcal{P}_+(I)$ such that $\mu(0) = \mu$ and $\dot{\mu}(0) = v = X(\mu)$. We have

$$X\mathfrak{g}_\mu(Y, Z) = \frac{d}{dt}\bigg|_{t=0} \mathfrak{g}_{\mu(t)}(Y, Z)$$
$$= \frac{d}{dt}\bigg|_{t=0} \sum_{i \in I} \frac{Y_i Z_i}{\mu_i + t v_i}$$
$$= -\sum_{i \in I} \frac{v_i Y_i Z_i}{\mu_i^2} = -\sum_{i \in I} \frac{X_i Y_i Z_i}{\mu_i^2}.$$

Similarly, one obtains formulae for $Y\mathfrak{g}(X, Z)$ and $Z\mathfrak{g}(X, Y)$. Now replacing the above results in Eq. 6.18, we get

$$\mathfrak{g}_\mu(\nabla_X Y, Z)$$
$$= \frac{1}{2}\left\{ -\sum_{i \in I} \frac{X_i Y_i Z_i}{\mu_i^2} - \sum_{i \in I} \frac{X_i Y_i Z_i}{\mu_i^2} + \sum_{i \in I} \frac{X_i Y_i Z_i}{\mu_i^2} \right\}$$
$$= -\frac{1}{2}\sum_{i \in I} \frac{X_i Y_i Z_i}{\mu_i^2}. \tag{6.19}$$

On the other hand, we have

$$\sum_{i \in I} \mathfrak{g}_\mu(X, Y) Z_i = \mathfrak{g}_\mu(X, Y) \sum_{i \in I} Z_i = 0, \tag{6.20}$$

since Z is a constant vector field on $\mathcal{P}_+(I)$. Then Eq. 6.19 can be written as

$$\mathfrak{g}_\mu(\nabla_X Y, Z)$$
$$= -\frac{1}{2}\sum_{i \in I} \left(\frac{X_i Y_i}{\mu_i^2} - \mathfrak{g}_\mu(X, Y) \right) \mu_i \frac{Z_i}{\mu_i}$$
$$= \mathfrak{g}_\mu \left(-\frac{1}{2}\left(\frac{dX}{d\mu}\frac{dY}{d\mu} - \mathfrak{g}_\mu(X, Y) \right) \mu, Z \right).$$

which completes the proof. □

6.2.3 Geodesics on $\mathcal{P}_+$

Theorem 6.2 *Let $\mu = \sum_{i \in I} \mu_i \delta^i$ be a probability measure in $\mathcal{P}_+(I)$ and $v \in T_\mu \mathcal{P}_+(I)$ a unit tangent vector, i.e., $||v||_\mu = 1$. Then the geodesic α that satisfies $\alpha(0) = \mu$ and $\dot{\alpha}(0) = v$ is given by $\alpha(t) = \sum_{i \in I} \alpha_i(t) \delta^i$ with*

6.2 Geometry of the Space of Probability Measures

$$\alpha_i(t) = \left(\cos\frac{t}{2} + \frac{\dot{\alpha}_i(0)}{\alpha_i(0)}\sin\frac{t}{2}\right)^2 \alpha_i(0), \qquad (6.21)$$

where $\alpha_i(0) = \mu_i$ and $\dot{\alpha}_i(0) = v_i$, $\forall i \in I$.

Proof Let $\alpha(t) = \sum_{i \in I} \alpha_i(t)\delta^i$ and $\dot{\alpha}(t) = \sum_{i \in I} \dot{\alpha}_i(t)\delta^i$. Then for each t, we have

$$\begin{cases} \sum_{i \in I} \alpha_i(t) = 1, \text{ and } \alpha_i(t) > 0, \forall i \in I, \\ \sum_{i \in I} \dot{\alpha}_i(t) = 0. \end{cases} \qquad (6.22)$$

Set X a constant vector field in $\mathcal{P}_+(I)$. From the condition (6.10) of Levi-Civita connection, we have

$$\mathfrak{g}_{\alpha(t)}(\nabla_{\dot{\alpha}(t)}\dot{\alpha}(t), X) = \dot{\alpha}(t)\left(\mathfrak{g}_{\alpha(t)}(\dot{\alpha}(t), X)\right) - \mathfrak{g}_{\alpha(t)}(\dot{\alpha}(t), \nabla_{\dot{\alpha}(t)}X). \qquad (6.23)$$

With the properties of Levi-Civita connection, to compute $\nabla_{\dot{\alpha}(t)}X$, the tangent vector $\dot{\alpha}(t)$ can be considered as a constant vector field on $\mathcal{P}_+(I)$ when t is fixed. Therefore, applying (6.17) for $\dot{\alpha}(t)$ and X we get,

$$\begin{aligned}
\nabla_{\dot{\alpha}(t)}X &= -\frac{1}{2}\left(\frac{d\dot{\alpha}(t)}{d\alpha(t)}\frac{dX}{d\alpha(t)} - \mathfrak{g}_{\alpha(t)}(\dot{\alpha}(t), X)\right)\alpha(t) \\
&= -\frac{1}{2}\sum_{i \in I}\left(\frac{\dot{\alpha}_i}{\alpha_i}\frac{X_i}{\alpha_i} - \sum_{j \in I}\frac{\dot{\alpha}_j X_j}{\alpha_j}\right)\alpha_i\delta^i.
\end{aligned} \qquad (6.24)$$

Taking into account of Eq. 6.22, the last term in Eq. 6.23 becomes

$$\begin{aligned}
&\mathfrak{g}(\dot{\alpha}(t), \nabla_{\dot{\alpha}(t)}X) \\
&= \left\langle \frac{d\dot{\alpha}}{d\alpha}, \frac{d\nabla_{\dot{\alpha}(t)}X}{d\alpha}\right\rangle_{\alpha(t)} \\
&= -\frac{1}{2}\sum_{i \in I}\frac{\dot{\alpha}_i}{\alpha_i}\left(\frac{\dot{\alpha}_i}{\alpha_i}\frac{X_i}{\alpha_i} - \sum_{j \in I}\frac{\dot{\alpha}_j X_j}{\alpha_j}\right)\alpha_i \\
&= -\frac{1}{2}\sum_{i \in I}\frac{\dot{\alpha}_i^2 X_i}{\alpha_i^2}.
\end{aligned} \qquad (6.25)$$

Now, we compute the second term in Eq. 6.23. We have

$$\dot{\alpha}(t)\left(\mathfrak{g}_{\alpha(t)}(\dot{\alpha}(t), X)\right) = \frac{d}{dt}\mathfrak{g}_{\alpha(t)}(\dot{\alpha}(t), X) = \sum_{i \in I}\frac{d}{dt}\left(\frac{\dot{\alpha}_i}{\alpha_i}\right)X_i. \qquad (6.26)$$

Combining Eqs. 6.25 and 6.26 in Eq. 6.23, we get

$$\mathfrak{g}_{\alpha(t)}(\nabla_{\dot{\alpha}(t)}\dot{\alpha}(t), X) = \sum_{i \in I} \left(\frac{d}{dt}\left(\frac{\dot{\alpha}_i}{\alpha_i}\right) + \frac{1}{2}\frac{\dot{\alpha}_i^2}{\alpha_i^2} \right) X_i. \tag{6.27}$$

Let's define the function $F(t)$ as

$$F(t) = -\sum_{i \in I} \left(\frac{d}{dt}\left(\frac{\dot{\alpha}_i}{\alpha_i}\right) + \frac{1}{2}\frac{\dot{\alpha}_i^2}{\alpha_i^2} \right) \alpha_i(t)$$

$$= -\sum_{i \in I} \frac{d}{dt}\left(\frac{\dot{\alpha}_i}{\alpha_i}\right) \alpha_i(t) - \frac{1}{2}\mathfrak{g}_{\alpha(t)}(\dot{\alpha}(t), \dot{\alpha}(t)). \tag{6.28}$$

Hence, the measure

$$\nu(t) = \sum_{i \in I} \left(\frac{d}{dt}\left(\frac{\dot{\alpha}_i}{\alpha_i}\right) + \frac{1}{2}\frac{\dot{\alpha}_i^2}{\alpha_i^2} + F(t) \right) \alpha_i \delta^i \tag{6.29}$$

belongs to $T_{\alpha(t)}\mathcal{P}_+$. In this way, Eq. 6.27 can be written as $\mathfrak{g}_{\alpha}(\nabla_{\dot{\alpha}}\dot{\alpha}, X) = \mathfrak{g}_{\alpha}(\nu, X)$. Since X is an arbitrary constant vector field, we get

$$\nabla_{\dot{\alpha}}\dot{\alpha} = \nu = \sum_{i \in I} \left(\frac{d}{dt}\left(\frac{\dot{\alpha}_i}{\alpha_i}\right) + \frac{1}{2}\frac{\dot{\alpha}_i^2}{\alpha_i^2} + F(t) \right) \alpha_i \delta^i. \tag{6.30}$$

Therefore, $\alpha(t) = \sum_{i \in I} \alpha_i(t) \delta^i$ is a geodesic if and only if

$$\begin{cases} \frac{d}{dt}\left(\frac{\dot{\alpha}_i}{\alpha_i}\right) + \frac{1}{2}\left(\frac{\dot{\alpha}_i}{\alpha_i}\right)^2 + F(t) = 0, & \forall i \in I, \\ \sum_{i \in I} \dot{\alpha}_i(t) = 0, & \forall t. \end{cases} \tag{6.31}$$

The next goal is to solve Eq. 6.31. We may remark that if α is a geodesic then $\mathfrak{g}_{\alpha(t)}(\dot{\alpha}(t), \dot{\alpha}(t))$ is constant along $\alpha(t)$. Consequently, taking into account of the assumption that $||\dot{\gamma}(0)||_\mu = 1$, we can assert that

$$\mathfrak{g}_{\alpha(t)}(\dot{\alpha}(t), \dot{\alpha}(t)) = \sum_{i \in I} \frac{\dot{\alpha}_i^2}{\alpha_i} \equiv 1. \tag{6.32}$$

Thus

$$\sum_{i \in I} \frac{d}{dt}\left(\frac{\dot{\alpha}_i}{\alpha_i}\right) \alpha_i = \frac{d}{dt} \sum_{i \in I} \left(\frac{\dot{\alpha}_i}{\alpha_i}\alpha_i\right) - \sum_{i \in I} \frac{\dot{\alpha}_i^2}{\alpha_i} = -1. \tag{6.33}$$

Which gives that $F(t) = \frac{1}{2}$. Substituting this result in Eq. 6.31, we obtain

6.2 Geometry of the Space of Probability Measures

$$\frac{d}{dt}\left(\frac{\dot{\alpha}_i}{\alpha_i}\right) + \frac{1}{2}\left(\frac{\dot{\alpha}_i}{\alpha_i}\right)^2 + \frac{1}{2} = 0, \quad \forall i \in I. \tag{6.34}$$

Set $\omega_i(t) = \dfrac{\dot{\alpha}_i(t)}{\alpha_i(t)}$. Equation 6.34 is written as

$$\frac{d}{dt}\omega_i + \frac{1}{2}\omega_i^2 + \frac{1}{2} = 0, \quad \forall i \in I,$$

The solution of this differential equation is given by $\omega_i = \tan\left(-\frac{t}{2} + \Theta^i\right)$, where Θ^i is constant, $i \in I$. Hence, we have

$$\frac{\dot{\alpha}_i}{\alpha_i} = \tan\left(-\frac{1}{2}t + \Theta^i\right), \forall i \in I$$

and $\alpha_i(t) = \Omega^i \cos^2\left(-\frac{t}{2} + \Theta^i\right)$, where Ω^i is constant, and $i \in I$. Taking into account initial conditions, we find that

$$\Theta^i = \arctan\left(\frac{\dot{\alpha}_i(0)}{\alpha_i(0)}\right), \tag{6.35}$$

$$\Omega^i = \frac{\alpha_i^2(0) + \dot{\alpha}_i^2(0)}{\alpha_i(0)}. \tag{6.36}$$

which proves the theorem. □

Corollary 6.1 *The geodesic $\alpha(t)$ with $\alpha(0) = \mu$ and $\dot{\alpha}(0) = v$, where v is a nontrivial tangent vector (not necessary unit), is given by*

$$\alpha(t) = \sum_{i \in I}\left(\cos\frac{t\|v\|_\mu}{2} + \frac{v_i}{\mu_i\|v\|_\mu}\sin\frac{t\|v\|_\mu}{2}\right)^2 \mu_i \delta^i. \tag{6.37}$$

Proposition 6.2 *The Fisher–Rao distance $d^{FR} : \mathcal{P}_+(I) \times \mathcal{P}_+(I) \to [0, \pi)$ between two measures $\mu, \nu \in \mathcal{P}_+(I)$ under the Fisher–Rao metric is given by*

$$d^{FR}(\mu, \nu) = 2\arccos\left(\sum_{i \in I}\sqrt{\mu_i \nu_i}\right). \tag{6.38}$$

To prove Proposition 6.2, we will show the following lemma given in [16].

Lemma 6.1 *Let*

$$\mathbb{S}^+_{(0,2)}(I) = \left\{f \in \mathcal{F}(I) \mid f^i > 0, \forall i \in I \text{ and } \sum_{i \in I}(f^i)^2 = 4\right\}$$

be the positive sector of the sphere centered at 0 with radius 2. As a submanifold of $\mathcal{F}(I)$ it carries the induced standard metric of $\mathcal{F}(I)$. That is for a given point $f \in \mathbb{S}^+_{(0,2)}(I)$ and two tangents vectors $p, q \in T_f \mathbb{S}^+_{(0,2)}(I)$, we have

$$\langle p, q \rangle_f = \sum_{i \in I} p^i q^i. \tag{6.39}$$

Then the map $\Phi : \mathcal{P}_+(I) \longrightarrow \mathbb{S}^+_{(0,2)}(I)$ defined by

$$\mu = \sum_{i \in I} \mu_i \delta^i \longmapsto 2 \sum_{i \in I} \sqrt{\mu_i} e_i$$

is an isometry.

Proof of the Lemma It is clear that Φ is bijective. Now, let v, w be in $T_\mu \mathcal{P}_+(I)$. We have

$$\left\langle \frac{\partial \Phi}{\partial v}(\mu), \frac{\partial \Phi}{\partial w}(\mu) \right\rangle$$

$$= \left\langle \frac{d}{dt} \Phi(\mu + vt) \Big|_{t=0}, \frac{d}{dt} \Phi(\mu + wt) \Big|_{t=0} \right\rangle$$

$$= \left\langle \sum_{i \in I} \frac{v_i}{\sqrt{\mu_i}} e_i, \sum_{i \in I} \frac{w_i}{\sqrt{\mu_i}} e_i \right\rangle$$

$$= \sum_{i \in I} \frac{v_i w_i}{\mu_i} = \mathfrak{g}_\mu(v, w).$$

$\square$

Proof of the Proposition By virtue of Lemma 6.1, we get

$$d^{FR}(\mu, \nu) = d(\Phi(\mu), \Phi(\nu)) = 2 \arccos \left(\sum_{i \in I} \sqrt{\mu_i \nu_i} \right).$$

$\square$

Theorem 6.3 *Let μ, ν be two different probability measures in $\mathcal{P}_+(I)$. Then there exists a unique geodesic $\alpha : [0, l] \to \mathcal{P}_+(I), t \to \alpha(t)$, joining two points μ and ν, with $\alpha(0) = \mu, \alpha(l) = \nu$ and $l = d^{FR}(\mu, \nu)$, given by*

$$\alpha(t) = \sum_{i \in I} \left(\cos \frac{t}{2} + \frac{d\tau}{d\mu}(i) \sin \frac{t}{2} \right)^2 \mu_i \delta^i, \tag{6.40}$$

6.2 Geometry of the Space of Probability Measures

where τ is the tangent vector in $T_\mu \mathcal{P}_+(I)$ defined by

$$\tau = \frac{1}{\sin \frac{l}{2}} \sum_{i \in I} \left(\sqrt{\frac{dv}{d\mu}} - \sum_{j \in I} \sqrt{\frac{dv}{d\mu}(j)} \mu(j) \right) \mu_i \delta^i. \tag{6.41}$$

Proof The proof falls naturally into three parts.

Step 1 First, let us check that τ is a tangent vector in $T_\mu \mathcal{P}_+(I)$. Indeed,

$$\frac{1}{\sin \frac{l}{2}} \sum_{i \in I} \left(\sqrt{\frac{dv}{d\mu}(i)} - \sum_{j \in I} \sqrt{\frac{dv}{d\mu}(j)} \mu(j) \right) \mu_i$$

$$= \frac{1}{\sin \frac{l}{2}} \left(\sum_{i \in I} \sqrt{\frac{dv}{d\mu}(i)} \mu_i - \sum_{j \in I} \sqrt{\frac{dv}{d\mu}(j)} \mu(j) \right)$$

$$= 0. \tag{6.42}$$

Then, since

$$\left(\sum_{j \in I} \sqrt{\frac{dv}{d\mu}(j)} \mu(j) \right)^2 = \left(\sum_{j \in I} \sqrt{\mu_j v_j} \right)^2 = \cos^2 \frac{l}{2}. \tag{6.43}$$

it follows that

$$\langle \tau, \tau \rangle_\mu = \frac{1}{\sin^2 \frac{l}{2}} \sum_{i \in I} \left(\sqrt{\frac{dv}{d\mu}(i)} - \sum_{j \in I} \sqrt{\frac{dv}{d\mu}(j)} \mu(j) \right)^2 \mu_i$$

$$= \frac{1}{\sin^2 \frac{l}{2}} \left(\sum_{i \in I} v(i) - \left(\sum_{j \in I} \sqrt{\frac{dv}{d\mu}(j)} \mu(j) \right)^2 \right)$$

$$= \frac{1}{\sin^2 \frac{l}{2}} \left(1 - \cos^2 \frac{l}{2} \right) = 1. \tag{6.44}$$

hence τ is a unit tangent vector.

Step 2 Now let us examine that the curve $\alpha(t)$ defined in Eq. 6.40 satisfies $\alpha(0) = \mu$ and $\alpha(1) = v$. It is easily seen that for $t = 0$, $\alpha(0) = \mu$. Now for $t = l$, we have

$$\alpha(l) = \sum_{i \in I} \left(\cos \frac{l}{2} + \frac{d\tau}{d\mu}(i) \sin \frac{l}{2} \right)^2 \mu_i \delta^i, \tag{6.45}$$

By Eq. 6.41, we get:

$$\frac{d\tau}{d\mu} \sin \frac{l}{2} = \sum_{i \in I} \left(\sqrt{\frac{dv}{d\mu}}(i) - \sum_{j \in I} \sqrt{\frac{dv}{d\mu}}(j)\mu(j) \right) e_i$$

$$= \sum_{i \in I} \left(\sqrt{\frac{dv}{d\mu}}(i) - \cos \frac{l}{2} \right) e_i. \tag{6.46}$$

Hence,

$$\alpha(l) = \sum_{i \in I} \left(\cos \frac{l}{2} + \sqrt{\frac{dv}{d\mu}}(i) - \cos \frac{l}{2} \right)^2 \mu_i \delta^i$$

$$= \sum_{i \in I} v_i \delta^i = v. \tag{6.47}$$

Step 3 Now we go to prove the uniqueness of the curve. Let $\mu(t) = \exp_\mu \tau t$ and $\tilde{\mu}(t) = \exp_\mu \tilde{\tau} t$ be unit speed geodesics corresponding with τ and $\tilde{\tau}$, and satisfying $\mu(0) = \tilde{\mu}(0) = \mu$ and $\mu(l) = \tilde{\mu}(l) = v$. By means of Theorem 6.2, we have

$$\mu(t) = \sum_{i \in I} \left(\cos \frac{t}{2} + \frac{d\tau}{d\mu} \sin \frac{t}{2} \right)^2 \mu_i \delta^i, \tag{6.48}$$

$$\tilde{\mu}(t) = \sum_{i \in I} \left(\cos \frac{t}{2} + \frac{d\tilde{\tau}}{d\mu} \sin \frac{t}{2} \right)^2 \mu_i \delta^i. \tag{6.49}$$

From later condition, we have

$$\left(\cos \frac{l}{2} + \frac{d\tau}{d\mu}(i) \sin \frac{l}{2} \right)^2 = \left(\cos \frac{l}{2} + \frac{d\tilde{\tau}}{d\mu}(i) \sin \frac{l}{2} \right)^2, \forall i \in I \tag{6.50}$$

$$\Rightarrow \cos \frac{l}{2} + \frac{d\tau}{d\mu}(i) \sin \frac{l}{2} = \pm \left(\cos \frac{l}{2} + \frac{d\tilde{\tau}}{d\mu}(i) \sin \frac{l}{2} \right), \forall i \in I. \tag{6.51}$$

Define

$$I_\pm = \left\{ i \in I \,\middle|\, \cos \frac{l}{2} + \frac{d\tau}{d\mu}(i) \sin \frac{l}{2} \right.$$
$$= \pm \left(\cos \frac{l}{2} + \frac{d\tilde{\tau}}{d\mu}(i) \sin \frac{l}{2} \right) \right\} \tag{6.52}$$

Then we have $I_- \cup I_+ = I$. Moreover $I_- \cap I_+ = \emptyset$. Indeed, if there exists $i \in I_- \cap I_+$ then

6.2 Geometry of the Space of Probability Measures

$$v_i = \left(\cos\frac{t}{2} + \frac{d\tau}{d\mu}\sin\frac{t}{2}\right)^2 \mu_i = 0, \quad (6.53)$$

contradict to $v \in \mathcal{P}_+$. Sine $0 < l < \pi$, we have

$$I_+ = \{i \in I | \tau_i = \tilde{\tau}_i\}, \quad (6.54)$$

$$I_- = \left\{i \in I | \tau_i + \tilde{\tau}_i = -2\mu_i \cot\frac{l}{2}\right\}. \quad (6.55)$$

Suppose $I_- \neq \emptyset$, since τ and $\tilde{\tau}$ are unit tangent vectors at μ, we have

$$\sum_{i \in I_+} \tau_i + \sum_{i \in I_-} \tau_i = \sum_{i \in I_+} \tilde{\tau}_i + \sum_{i \in I_-} \tilde{\tau}_i = 0 \quad (6.56)$$

$$\Rightarrow \sum_{i \in I_-}\left(\tilde{\tau}_i + 2\mu_i \cot\frac{l}{2}\right) + \sum_{i \in I_-} \tilde{\tau}_i = 0. \quad (6.57)$$

Since Eq. 6.57 we see that if $I_- = I$, then $\cot\frac{l}{2} = 0$ contradicts to $0 < l < \pi$. So $I_- \neq I$. We have the claim below. $\square$

Claim 6.1 *For all $\mu \in \mathcal{P}_+$ and $0 < l < \pi$. If $\tau, \tilde{\tau} \in T_\mu \mathcal{P}_+$. Let*

$$I_+ = \{i \in I | \tau_i = \tilde{\tau}_i\}, \quad (6.58)$$

$$I_- = \left\{i \in I | \tau_i + \tilde{\tau}_i = -2\mu_i \cot\frac{l}{2}\right\}. \quad (6.59)$$

then $I_- = \emptyset$.

Proof of the Claim We proof the Claim by induction on the degree of I. If $|I|$ is one or two the Claim is true since I_+ is not empty. Suppose the Claim is true for $|I| = n$. We go to prove the Claim for $|I| = n + 1$. Let $\mu, v, \tilde{v}$ and l as in the Claim. Suppose $I_- \neq \emptyset$ then $|I_-| \geq 2$. Let g, h be two distinct index in I_-, this means $v_g + \tilde{v}_g = -2\mu_g \cot\frac{l}{2}$ and $v_h + \tilde{v}_h = -2\mu_h \cot\frac{l}{2}$. Now let $k \in I_+$ and define three measures $v', \tilde{v}', \mu'$ on $I \setminus \{k\}$ as follow

$$v' = \sum_{i \in I, i \neq k, h, g} v_i \delta^i + v_g \delta^g + (v_h + v_k)\delta^h, \quad (6.60)$$

$$\tilde{v}' = \sum_{i \in I, i \neq k, h, g} \tilde{v}_i \delta^i + (\tilde{v}_g + 2\tilde{v}_k)\delta^g + (\tilde{v}_h - \tilde{v}_k)\delta^h, \quad (6.61)$$

$$\mu' = \sum_{i \in I, i \neq k, h, g} \mu_i \delta^i + (\mu_g + \mu_k)\delta^g + \mu_h \delta^h. \quad (6.62)$$

We have $v', \tilde{v}' \in T_{\mu'}\mathcal{P}_+(I \setminus \{k\})$, and $h \in I_- \neq \emptyset$. This contradicts to the hypothesis. This shows the Claim for $|I| = n + 1$. $\square$

By means of the Claim, we prove the uniqueness of the geodesic (6.40) defined with the unit tangent vector (6.41). □

Corollary 6.2 *Let*

$$\varepsilon = \{(\mu, v) \,|\, \alpha(t, \mu, v) \text{ is defined on an interval containing } [0, l]\,\}$$

The exponential map $\exp_\mu : \varepsilon \longrightarrow \mathcal{P}_+(I)$ *is defined as*

$$\exp_\mu(v) = \sum_{i \in I} \left(\cos \frac{\|v\|_\mu}{2} + \frac{v_i}{\mu_i \|v\|_\mu} \sin \frac{\|v\|_\mu}{2} \right)^2 \mu_i \delta^i. \tag{6.63}$$

Similarly, given two points μ and v on $\mathcal{P}_+(I)$, the inverse exponential map (also known as the logarithmic map) at μ, $\log_\mu : \mathcal{P}_+(I) \longrightarrow \varepsilon$ is defined for any $v \in \mathcal{P}_+(I)$ by

$$\log_\mu(v) = \frac{l}{\sin \frac{l}{2}} \sum_{i \in I} \left(\sqrt{\frac{dv}{d\mu}}(i) - \sum_{j \in I} \sqrt{\frac{dv}{d\mu}}(j) \mu(j) \right) \mu_i \delta^i. \tag{6.64}$$

Theorem 6.4 $\mathcal{P}_+(I)$ *equipped with the Fisher–Rao metric is a locally symmetric space.*

To show that $\mathcal{P}_+(I)$ is a locally symmetric space, we need the following proposition.

Proposition 6.3 *A locally symmetric space is a Riemannian manifold whose curvature tensor is parallel, i.e., $\nabla R = 0$.*

Proof of Theorem 6.4 It follows by Lemma 6.1 as $\mathcal{P}_+(I)$ is isometric to the symmetric space $\mathbb{S}_{(0,2),+}(I)$. Hence, around any point $\mu \in \mathcal{P}_+(I)$ there is a geodesic ball $B_r(\mu)$ which is isometric to a ball $B_\epsilon(f)$ on $\mathbb{S}_{(0,2),+}(I)$. Therefore, there is an isometry $\kappa_\mu : B_r(\mu) \to B_r(f)$ such that $\kappa_\mu(\mu) = \mu$ and the differential $d\kappa_\mu(\mu) = -Id$. Let $v^{(1)}, v^{(2)}, v^{(3)}, v^{(4)} \in T_\mu(\mathcal{P}_+(I))$. Let us show that $\nabla_{v^{(1)}} R(v^{(2)}, v^{(3)}, v^{(4)}) = 0$. Set $v = \nabla_{v^{(1)}} R(v^{(2)}, v^{(3)}, v^{(4)})$ and applying $d\kappa_\mu$ on both sides of this equation, we obtain that v will be changed by $-v$ and $v^{(i)}$ to $-v^{(i)}$, for $i = 1, \ldots, 4$. Thus the left-hand side changes sign while the right-hand side stays the same. Moreover, we mention that the curvature tensor is preserved by $d\kappa_\mu$. Consequently, $\nabla R = 0$ and the proof is complete. □

6.2.4 Levi-Civita Parallel Transport on $\mathcal{P}_+$

Let us consider two points $\mu, v \in \mathcal{P}_+(I)$, a tangent vector $v \in T_\mu \mathcal{P}_+(I)$ and a geodesic curve $\alpha : [0, l] \to \mathcal{P}_+(I)$ on $\mathcal{P}_+(I)$ such that $\alpha(0) = \mu$ and $\alpha(l) = v$. We would like to map v from $T_\mu \mathcal{P}_+(I) = T_{\alpha(0)} \mathcal{P}_+(I)$ to $T_v \mathcal{P}_+(I) = T_{\alpha(l)} \mathcal{P}_+(I)$. We

6.2 Geometry of the Space of Probability Measures

introduce X, a vector field defined along the geodesic α, such that $X(\mu) = v$ and $\nabla_{\dot{\alpha}(t)} X(\alpha(t)) = 0$. We say that the tangent vector v is constant along the geodesic curve α with respect to ∇.

Definition 6.4 A metric parallel transport on $\mathcal{P}_+(I)$ is the map

$$\Gamma_{\alpha(0) \to \alpha(t)} : T_{\alpha(0)}\mathcal{P}_+(I) \to T_{\alpha(t)}\mathcal{P}_+(I) \tag{6.65}$$

such that for any $v, w \in T_\mu \mathcal{P}_+(I)$, and for $t \in [0, l]$ we have

$$\mathfrak{g}_{\alpha(0)}(v, w) = \mathfrak{g}_{\alpha(t)} \left(\Gamma_{\alpha(0) \to \alpha(t)}(v), \Gamma_{\alpha(0) \to \alpha(t)}(w) \right). \tag{6.66}$$

Γ is the Levi-Civita parallel transport along the geodesic curve α on $\mathcal{P}_+(I)$ with respect to the Fisher–Rao metric.

Rewriting equation $\nabla_{\dot{\alpha}(t)} X(\alpha(t)) = 0$, we conclude that computing $X(t) = X(\alpha(t))$ requires solving a linear first order differential equations on $\mathcal{P}_+(I)$ given by

$$\frac{dX_k}{dt} + \sum_{i,j} \alpha_{ij}^k \frac{d\alpha_i}{dt} X_j = 0, \quad \text{for } k = 1, \ldots, n. \tag{6.67}$$

We check at once that it is difficult to solve Eq. 6.67 directly. Hence we will use Eq. 6.17.

Theorem 6.5 *Let μ be a probability measure in $\mathcal{P}_+(I)$ and $v \in T_\mu \mathcal{P}_+(I)$ a unit tangent vector, i.e., $\|v\|_\mu = 1$. Let $\alpha : [0, l] \to \mathcal{P}_+(I)$ be a geodesic curve such that $\alpha(0) = \mu$ and $\dot{\alpha}(0) = v$. The Levi-Civita parallel transport of a vector $w \in T_\mu \mathcal{P}_+(I)$ to $T_{\alpha(t)}\mathcal{P}_+(I)$, is given by*

$$\Gamma_{\alpha(0) \to \alpha(t)}(w) = \sum_{i \in I} \sqrt{\alpha_i(t)} \left(-F(0)\sqrt{\mu_i} \left(2 \sin \frac{t}{2} - 2\frac{v_i}{\mu_i} \cos \frac{t}{2} \right) \right.$$
$$\left. + \frac{w_i}{\sqrt{\mu_i}} - 2F(0)\frac{v_i}{\sqrt{\mu_i}} \right) \delta^i, \tag{6.68}$$

where $F(0) = \frac{1}{2}\mathfrak{g}_\mu(v, w)$.

Proof We can proceed analogously to the proof of Theorem 6.2. Thus, let $\alpha(t) = \sum_{i \in I} \alpha_i(t)\delta^i$ be a geodesic curve, and define $\dot{\alpha}(t) = \sum_{i \in I} \dot{\alpha}_i(t)\delta^i$. Consider the vector field X on α defined by $X(\alpha(t)) = \sum_{i \in I} X_i(\alpha(t))\delta^i$, for $t \in [0, l]$, as the parallel transport of vector w along α. Then

$$\begin{cases} \nabla_{\dot{\alpha}(t)} X(t) = 0 \\ X(0) = w \end{cases}, \tag{6.69}$$

where we write $X(\alpha(t))$ simply $X(t)$ when no confusion can arise. Let Y be a constant vector field (in the sense of Definition 6.3) on $\mathcal{P}_+(I)$, we have

$$\mathfrak{g}_{\alpha(t)}\left(\nabla_{\dot{\alpha}(t)}X(t), Y\right) = \dot{\alpha}(t)\left(\mathfrak{g}_{\alpha(t)}(X(t), Y)\right) - \mathfrak{g}_{\alpha(t)}\left(X(t), \nabla_{\dot{\alpha}(t)}Y\right). \tag{6.70}$$

Applying Theorem 7.2, we get

$$\nabla_{\dot{\alpha}}Y = -\frac{1}{2}\sum_{i\in I}\left(\frac{\dot{\alpha}_i\, Y_i}{\alpha_i\, \gamma_i} - \sum_{j\in I}\frac{\dot{\alpha}_j Y_j}{\alpha_j}\right)\alpha_i \delta^i. \tag{6.71}$$

Hence the last term in Eq. 6.70 becomes

$$\mathfrak{g}_{\alpha}(X, \nabla_{\dot{\alpha}}Y) = -\frac{1}{2}\sum_{i\in I}\frac{X_i}{\alpha_i}\left(\frac{\dot{\alpha}_i\, Y_i}{\alpha_i\, \alpha_i} - \sum_{j\in I}\frac{\dot{\alpha}_j Y_j}{\alpha_j}\right)\alpha_i$$

$$= -\frac{1}{2}\sum_{i\in I}\frac{X_i Y_i \dot{\alpha}_i}{\alpha_i^2}. \tag{6.72}$$

Let us now compute the second term in Eq. 6.70. We obtain

$$\dot{\alpha}(t)\left(\mathfrak{g}_{\alpha(t)}(X, Y)\right) = \frac{d}{dt}\mathfrak{g}_{\alpha(t)}(X(t), Y)$$

$$= \sum_{i\in I}\frac{d}{dt}\left(\frac{X_i}{\alpha_i}\right)Y_i. \tag{6.73}$$

Consequently, Eq. 6.70 becomes

$$\mathfrak{g}_{\alpha}(\nabla_{\dot{\alpha}}X, Y) = \sum_{i\in I}\left(\frac{d}{dt}\left(\frac{X_i}{\alpha_i}\right) + \frac{1}{2}\frac{X_i \dot{\alpha}_i}{\alpha_i^2}\right)Y_i. \tag{6.74}$$

Define the function $F(t)$ by

$$F(t) = -\sum_{i\in I}\left(\frac{d}{dt}\left(\frac{X_i}{\alpha_i}\right) + \frac{1}{2}\frac{X_i \dot{\alpha}_i}{\alpha_i^2}\right)\alpha_i(t)$$

$$= -\sum_{i\in I}\frac{d}{dt}\left(\frac{X_i}{\alpha_i}\right)\alpha_i(t) - \frac{1}{2}\mathfrak{g}_{\alpha(t)}(X(t), \dot{\alpha}(t)). \tag{6.75}$$

6.2 Geometry of the Space of Probability Measures

Then, $\forall t \in [0, l]$, the probability measure

$$v(t) = \sum_{i \in I} \left(\frac{d}{dt}\left(\frac{X_i}{\alpha_i}\right) + \frac{1}{2}\frac{X_i \dot{\alpha}_i}{\alpha_i^2} + F(t) \right) \alpha_i \delta^i$$

belongs to $T_{\alpha(t)}\mathcal{P}_+(I)$. Thus, Eq. 6.74 can be written as

$$\mathfrak{g}_\alpha(\nabla_{\dot{\alpha}} X, Y) = \mathfrak{g}_\alpha(v, Y). \tag{6.76}$$

Since Y is an arbitrary constant vector field, we get

$$\nabla_{\dot{\alpha}} X = v = \sum_{i \in I} \left(\frac{d}{dt}\left(\frac{X_i}{\alpha_i}\right) + \frac{1}{2}\frac{X_i \dot{\alpha}_i}{\alpha_i^2} + F(t) \right) \alpha_i \delta^i. \tag{6.77}$$

Therefore, $X(t)$ is the parallel transport of the vector w along the geodesic curve $\alpha(t)$ if and only if

$$\begin{cases} \dfrac{d}{dt}\left(\dfrac{X_i}{\alpha_i}\right) + \dfrac{1}{2}\dfrac{X_i \dot{\alpha}_i}{\alpha_i^2} + F(t) = 0, & \forall i \in I, \\ X(0) = w. \end{cases} \tag{6.78}$$

The next concern will be to solve Eq. 6.78. We remind that

$$\mathfrak{g}_{\alpha(t)}(X(t), \dot{\alpha}(t)) = \mathfrak{g}_{\alpha(0)}(X(0), \dot{\alpha}(0)). \tag{6.79}$$

Moreover

$$\sum_{i \in I} \frac{d}{dt}\left(\frac{X_i}{\alpha_i}\right) \alpha_i = \frac{d}{dt}\sum_{i \in I}\left(\frac{X_i}{\alpha_i}\alpha_i\right) - \sum_{i \in I}\left(\frac{X_i \dot{\alpha}_i}{\alpha_i}\right)$$
$$= -\mathfrak{g}_{\alpha(0)}(X(0), \dot{\alpha}(0)). \tag{6.80}$$

Which gives that $F(t)$ is a constant function and $F(t) = F(0) = \frac{1}{2}\mathfrak{g}_{\alpha(0)}(X(0), \dot{\alpha}(0))$. Hence, substituting this result in Eq. 6.78 we get

$$\frac{d}{dt}\left(\frac{X_i}{\alpha_i}\right) + \frac{1}{2}\frac{X_i \dot{\alpha}_i}{\alpha_i^2} + F(0) = 0, \quad \forall i \in I. \tag{6.81}$$

Set $\omega_i = \dfrac{X_i}{\alpha_i}$. Equation 6.81 can be written as

$$\frac{d}{dt}\omega_i + \frac{1}{2}\frac{\dot{\alpha}_i}{\alpha_i}\omega_i + F(0) = 0, \quad \forall i \in I. \tag{6.82}$$

Solution of the first order differential Eq. 6.82 is given by

$$\omega_i(t) = \frac{1}{\sqrt{\alpha_i(t)}} \left(-F(0)\sqrt{\alpha_i(0)} \left(2\sin\frac{t}{2} - 2\frac{\dot{\alpha}_i(0)}{\alpha_i(0)} \cos\frac{t}{2} \right) \right. \tag{6.83}$$
$$\left. + \Theta_i \right) \quad \text{for } \Omega_i \text{ constant}, i \in I.$$

Therefore,

$$X_i = \sqrt{\alpha_i(t)} \left(-F(0)\sqrt{\alpha_i(0)} \left(2\sin\frac{t}{2} - 2\frac{\dot{\alpha}_i(0)}{\alpha_i(0)} \cos\frac{t}{2} \right) + \Theta \right), \tag{6.84}$$

for Θ_i constant, $i \in I$. According to the initial conditions, it follows that

$$\Theta = \frac{w_i}{\sqrt{\mu_i}} - 2F(0)\frac{v_i}{\sqrt{\mu_i}}. \tag{6.85}$$

We conclude that

$$X_i(t) = \sqrt{\alpha_i(t)} \left(-F(0)\sqrt{\mu_i} \left(2\sin\frac{t}{2} - 2\frac{v_i}{\mu_i} \cos\frac{t}{2} \right) \right.$$
$$\left. + \frac{w_i}{\sqrt{\mu_i}} - 2F(0)\frac{v_i}{\sqrt{\mu_i}} \right), \quad i \in I. \tag{6.86}$$

and it is easy to check that, $\forall t \in [0, l]$, $X(t) = \sum_{i \in I} X_i(t)\delta^i \in T_{\gamma(t)}\mathcal{P}_+(I)$ and it is the Levi-Civita parallel transport of the vector w along the geodesic curve $\alpha(t)$. □

Theorem 6.6 *Given two distinct probability measures μ and v in $\mathcal{P}_+(I)$, a nontrivial tangent vector $w \in T_\mu \mathcal{P}_+(I)$ and a geodesic curve $\alpha : [0, l] \to \mathcal{P}_+(I)$ such that $\alpha(0) = \mu$ and $\alpha(l) = v$. The Levi-Civita parallel transport, $\Gamma_{\mu \mapsto v} : T_\mu \mathcal{P}_+(I) \to T_v \mathcal{P}_+(I)$, that transports a vector w from $T_\mu \mathcal{P}_+(I) = T_{\alpha(0)}\mathcal{P}_+(I)$ to $T_v \mathcal{P}_+(I) = T_{\alpha(l)}\mathcal{P}_+(I)$ given by*

$$\Gamma_{\mu \mapsto v}(w) = \sum_{i \in I} \sqrt{v_i} \left(-F(0)\sqrt{\mu_i} \left(2\sin\frac{l}{2} - 2\frac{\tau_i}{\mu_i} \cos\frac{l}{2} \right) \right.$$
$$\left. + \frac{w_i}{\sqrt{\mu_i}} - 2F(0)\frac{\tau_i}{\sqrt{\mu_i}} \right)\delta^i, \tag{6.87}$$

where $l = 2\arccos\sum_{i \in I} \sqrt{\mu_i v_i}$, $F(0) = \frac{1}{2}\mathfrak{g}_\mu(w, \tau)$, and τ is the unit tangent vector

$$\tau = \frac{1}{\sin\frac{l}{2}} \sum_{i \in I} \left(\sqrt{\frac{dv}{d\mu}}(i) - \sum_{j \in I} \sqrt{\frac{dv}{d\mu}}(j)\mu(j) \right) \mu_i \delta^i. \tag{6.88}$$

6.3 Interpolation Problem on Space of Probability Measures

Table 6.1 Geometric toolbox for the space of probability measures $\mathcal{P}_+(I)$

Set	$\mathcal{P}_+(I) = \{\mu = \sum_{i \in I} \mu_i \delta^i \mid \mu_i > 0, \ \forall i \in I, \text{ and } \sum_{i \in I} \mu_i = 1\}$
Tangent spaces	$T_\mu \mathcal{P}_+(I) = \{(\mu, v) \mid \mu \in \mathcal{P}_+(I) \text{ and } v \in \mathcal{S}_0(I)\}$
Inner product	$\mathfrak{g}_\mu(v, w) = \sum_{i \in I} \frac{v_i w_i}{\mu_i}$
Geodesic distance	$d^{FR}(\mu, \nu) = 2 \arccos\left(\sum_{i \in I} \sqrt{\mu_i \nu_i}\right)$
Shortest geodesic	$\alpha(t) = \sum_{i \in I} \left(\cos \frac{t\|v\|_\mu}{2} + \frac{v_i}{\mu_i \|v\|_\mu} \sin \frac{t\|v\|_\mu}{2}\right)^2 \mu_i \delta^i$
Exponential	$\exp_\mu(v) = \sum_{i \in I} \left(\cos \frac{\|v\|_\mu}{2} + \frac{v_i}{\mu_i \|v\|_\mu} \sin \frac{\|v\|_\mu}{2}\right)^2 \mu_i \delta^i$
Logarithm	$\log_\mu(\nu) = \frac{l}{\sin \frac{l}{2}} \sum_{i \in I} \left(\sqrt{\frac{d\nu}{d\mu}}(i) - \sum_{j \in I} \sqrt{\frac{d\nu}{d\mu}}(j) \mu(j)\right) \mu_i \delta^i$

Proof It suffices to use the equation of the geodesic curve $\alpha(t)$ joining two points μ and ν given by Theorem 6.3 together with taking $t = l$ in Theorem 6.5, the proof follows. □

The geometric toolbox presented in Table 6.1 summarizes different geometric structures on $\mathcal{P}_+(I)$.

6.3 Interpolation Problem on Space of Probability Measures

Given $\mu^{(0)}, \ldots, \mu^{(N)}$, an indexed finite set of probability measures associated with a set of observation times $t_0, \ldots, t_N$ such that $t_i = i$ for $i = 0, \ldots, N$. The objective is to devise a method for smooth interpolation on the space of probability measures $\mathcal{P}_+(I)$. More precisely, we seek a spline $\gamma : [t_0, t_N] \to \mathcal{P}_+(I)$ that interpolates the given set of probability measures: $\gamma(t_i) = \mu_i$ for $i = 0, \ldots, N$ and is of class C^2. For simplicity, we may consider the time interval $[0, N]$ instead of $[t_0, t_N]$. Solutions to this problem will be referred to as measure interpolation splines and aim to minimize the total cost functional:

$$E(\gamma) = \frac{\lambda}{2} \int_{t_0}^{t_n} < \frac{D^2 \gamma(t)}{Dt^2}, \frac{D^2 \gamma(t)}{Dt^2} >_F dt$$

To address the challenge of defining and computing splines of measures, we propose a novel optimization method based on the generalized De Casteljau algorithm on manifolds. The implementation of the De Casteljau algorithm on $\mathcal{P}_+(I)$ serves as the foundation for obtaining solutions to the optimization problem (6.3) in the probability measures space $\mathcal{P}_+(I)$. Additionally, we will decompose the interpolation problem into two steps: firstly, solving a C^1 interpolation spline on $\mathcal{P}_+(I)$, and subsequently, considering the local symmetries at interpolation points to derive equations for control points that generate the desired C^2 Bézier spline. The effectiveness of the method is demonstrated through the design of several measure interpolation splines on $\mathcal{P}_+(I)$, showcasing their applicability in various scenarios.

6.3.1 Measure Interpolation Spline Using De Casteljau Algorithm

As explored in preceding chapters, the De Casteljau algorithm stands out as a theoretically robust method for generating polynomial spline curves in Euclidean spaces. Its efficacy lies in its straightforward geometric construction, relying on uncomplicated successive linear interpolation. Extending the De Casteljau algorithm to Riemannian manifolds is a natural progression, achieved by replacing straight lines with minimal geodesics between two points, as detailed in [17]. However, when dealing with the space of probability measures, which is not geodesically complete, there are situations where an explicit formula for the geodesic joining two points might be unknown. Consequently, the implementation of the De Casteljau algorithm is confined to a convex open subset of the manifold where the expression for computing the geodesic arc between two points is well-defined.

Henceforth, let us denote $\gamma_j : [0, 1] \to \mathcal{P}_+(I)$ as the Bézier curve of order j, characterized by $(j+1)$ control points $V^{(0)}, \ldots, V^{(j)} \in \mathcal{P}_+(I)$. Introducing $(V^{(i)})^0(t) = V^{(i)}$, for $i = 0, \ldots, j-k$, where $k = 1, \ldots, j$, we proceed to define:

$$(V^{(i)})^k(t) = \gamma_k(t, V^{(i)}, \ldots, V^{(i+k)}) = \alpha\left(t, (V^{(i)})^{k-1}, (V^{(i+1)})^{k-1}\right), t \in [0, 1],$$

where α is the geodesic curve on $\mathcal{P}_+(I)$ given by Eq. 6.37. In this section, we construct the Bézier spline $\gamma : [0, N] \to \mathcal{P}_+(I)$ as a sequence of N Bézier curves γ_j^i of orders two and three. Specifically, the first and the last curves are quadratic Bézier curves, while all intermediate ones are cubic. Bézier curves of order j represent polynomial functions parametrized by a set of control points. The first and last control points coincide with the endpoints of the curve, but the intermediate control points are generally not located on the curve. The degree j of the polynomial spline is determined by the number of control points. Let $((\widehat{\eta}^{(i)})^-, (\widehat{\eta}^{(i)})^+)$ denote the control points on the left and right-hand side of the interpolation point $\mu^{(i)}$ for $i = 1, \ldots, (N-1)$. The Bézier spline $\gamma : [0, N] \longrightarrow \mathcal{P}_+(I)$ is then expressed as follows:

$$\gamma(t) = \begin{cases} \gamma_2^0(t; \mu^{(0)}, (\widehat{\eta}^{(1)})^-, \mu^{(1)}), & 0 \leq t \leq 1, \\ \gamma_3^i(t-i; \mu^{(i)}, (\widehat{\eta}^{(i)})^+, (\widehat{\eta}^{(i+1)})^-, \mu^{(i+1)}), & i-1 \leq t \leq i \\ \gamma_2^{N-1}(t-(N-1); \mu^{(N-1)}, (\widehat{\eta}^{(N-1)})^+, \mu^{(N)}), & N-1 \leq t \leq N \end{cases}$$

The Bézier spline γ interpolates the first and last control points of each Bézier curve γ_j^i, where $j \in \{2, 3\}$ and $0 \leq i \leq N-1$. Consequently, the continuity of γ at joint points is inherently satisfied. To achieve C^1 continuity along the curve, we propose a strategy that involves shifting the problem to the tangent space $T_{\mu^{(i)}}\mathcal{P}_+(I)$

6.3 Interpolation Problem on Space of Probability Measures

for $i = 1, \ldots, N - 1$ and subsequently bringing back the solution to the space of probability measures $\mathcal{P}_+(I)$. The proposed algorithm relies solely on the Riemannian exponential and logarithm maps. In detail, given $\mu^{(0)}, \ldots, \mu^{(N)}$, an indexed set of probability measures on $\mathcal{P}_+(I)$, we use the Riemannian logarithmic map Eq. 6.64 to lift the data points $\mu^{(0)}, \ldots, \mu^{(N)}$ into each tangent space $T_{\mu^{(i)}}\mathcal{P}_+(I)$, where $i = 1, \ldots, N - 1$. The mapped data are represented by $\phi^i = ((\phi^{(0)})^i, \ldots, (\phi^{(N)})^i)$ with $(\phi^{(k)})^i = \text{Log}_{\mu^{(i)}}(\psi^{(k)})$ for $k = 0, \ldots, N$. Let $\beta : [0, N] \to T_{\mu^{(i)}}\mathcal{P}_+(I)$ denote the Bézier spline on $T_{\mu^{(i)}}\mathcal{P}_+(I)$, where $i = 1, \ldots, N - 1$, defined by n Bézier curves β_j^i with $j \in \{2, 3\}$ and $0 \leq i \leq N - 1$. Consequently, the optimization problem (6.3) can be reformulated on $T_{\mu_i}\mathcal{P}_+(I)$ as follows:

$$\min_{((b^{(1)})^i)^-, \ldots, ((b^{(N-1)})^i)^-} E((b^{(1)})^i)^-, \ldots, ((b^{(N-1)})^i)^-) \tag{6.89}$$

$$:= \min_{((b^{(1)})^i)^-, \ldots, ((b^{(N-1)})^i)^-} \int_0^1 \|(\ddot{\beta}_2^i)^0(t; (\phi^{(0)})^i, ((b^{(1)})^i)^-, (\phi^{(1)})^i)\|_2^2 \tag{6.90}$$

$$+ \sum_{i=1}^{N-2} \int_0^1 \|\ddot{\beta}_3^i(t; (\phi^{(i)})^i, ((b^{(i)})^i)^+, ((b^{(i+1)})^i)^-, (\phi^{(i+1)})^i)\|_2^2 \tag{6.91}$$

$$+ \int_0^1 \|\ddot{\beta}_2^{N-1}(t; (\phi^{(N-1)})^i, ((b^{(N-1)})^i)^+, (\phi^{(N)})^i)\|_2^2. \tag{6.92}$$

where $((b^{(1)})^i)^-$ and $((b^{(1)})^i)^+$ denote control points on the left and on the right-hand side of the interpolation point $(\phi^{(i)})^i$ for $i = 1, \ldots, (N - 1)$ on $T_{\mu^{(i)}}\mathcal{P}_+(I)$. Given that the tangent space is a vector space, we can approach problem (6.92) similarly to the Euclidean case $\mathbb{R}^n$. Specifically, we establish that solutions to the minimization problem for the mean square acceleration of the Bézier curve β correspond precisely to the control points of the curve. Additionally, we provide conditions under which the Bézier curve β attains C^2 smoothness. Detailed equations governing the control points of the C^2 Bézier spline on $\mathbb{R}^n$ can be found in Chap. 1. Finally, leveraging the Riemannian exponential map $\text{Exp}_{\mu^{(i)}}$ defined on $\mathcal{P}_+(I)$ by Eq. 6.63, we transport the control points $B^i = [((b^{(1)})^i)^-, \ldots, ((b^{(N-1)})^i)^-]^T$ of the C^2 Bézier curve β from the tangent space $T_{\mu^{(i)}}\mathcal{P}_+(I)$ to $\mathcal{P}_+(I)$. The resulting control points in $\mathcal{P}_+(I)$ are optimal, ensuring that the curve is of class C^1.

Theorem 6.7 *The Bèzier spline $\gamma : [0, N] \to \mathcal{P}_+(I)$ interpolating the data points $\mu^{(i)}$ on $\mathcal{P}_+(I)$ is of class C^1 and is uniquely defined by the set of control points $\widehat{\eta} = [(\widehat{\eta}^{(1)})^-, \ldots, (\widehat{\eta}^{(N-1)})^-]^T \in \mathbb{R}^{n(N-1) \times n}$ given by:*

$$(\widehat{\eta}^{(i)})^- = Exp_{\mu^{(i)}}(\tilde{b}^{(i)}), \; i = 1, \ldots, N - 1, \tag{6.93}$$

where $\tilde{b}_i$, represent the row i of B^i in $T_{\mu^{(i)}}\mathcal{P}_+(I)$, for $i = 1, \ldots, N - 1$.

Proof Similar to the Euclidean space $\mathbb{R}^n$, the differentiability condition at the interpolation points allows us to express control points $((b^{(1)})^i)^-$ in terms of $((b^{(1)})^i)^+$ as

$$((b^{(1)})^i)^- = (\phi^{(i)})^i + \lambda_i(((b^{(1)})^i)^+ - (\phi^{(i)})^i) \tag{6.94}$$

Considering that $\log_p(b) = b - p$ in the Euclidean case, hence the generalization of Eq. 6.94 on $\mathcal{P}_+(I)$ is given by

$$(\widehat{\eta}^{(i)})^+ = \mathrm{Exp}_{\mu^{(i)}}\left(\lambda_i \mathrm{Exp}_{\mu^{(i)}}^{-1}((\widehat{\eta}^i)^-)\right) \tag{6.95}$$

which assert the C^1 differentiability condition on $\mathcal{P}_+(I)$. □

Remark 6.2 The interpolation point $(\phi^{(N)})^i$ undergoes modification due to the C^2 differentiability condition of the curve β_i on $T_{\mu^{(i)}}\mathcal{P}_+(I)$, for $i = 1, \ldots, N-1$. Consequently, the point $\mu^{(N)}$ is adjusted, resulting in the determination of new $(N+1)$ interpolation points on $\mathcal{P}_+(I)$, which are expressed as:

$$\tilde{\mu}^{(k)} = \mathrm{Exp}_{\mu^{(i)}}((\tilde{\phi}^{(k)})^i), \ k = 0, \ldots, N; \ i = 1, \ldots, N-1, \tag{6.96}$$

where $\tilde{\phi}^i = [(\tilde{\phi}^{(0)})^i, \ldots, (\tilde{\phi}^{(N)})^i]^T$ a matrix of size $n(N+1) \times n$ containing the new $(N+1)$ interpolation points in each tangent space $T_{\mu^{(i)}}\mathcal{P}_+(I)$.

6.3.2 C^2 Bézier Splines on $\mathcal{P}_+$

The objective now is to demonstrate that Bézier curves $\gamma_j^i, j \in \{2, 3\}, i = 0, \ldots, N-1$, can be arranged in a manner that ensures C^2 differentiability at joint points, thus enabling γ to be a C^2 spline. The fundamental geometric element employed to achieve this objective is the existence of local symmetries at interpolation points within the Riemannian manifold $\mathcal{P}_+(I)$. Specifically, given that the space of probability measures is a locally symmetric space, at every point $\mu \in \mathcal{P}_+(I)$, there exists a local isometry φ defined on a neighborhood U of μ. This local isometry satisfies $\varphi(\mu) = \mu$ and has a differential given by $d\varphi_\mu = -Id$.

In line with the methodology applied in previous chapters, both Lemma 4.2 and Theorem 4.1 meticulously investigate the correlation between local symmetries at interpolation points and the C^2 differentiability of the Bézier spline within symmetric spaces. This valuable insight will once again be leveraged to determine the control points of the C^2 Bézier spline γ on $\mathcal{P}_+(I)$. Let $(\eta^{(i)})^-$ and $(\eta^{(i)})^+$ denote the new control points on the left and right sides of the interpolation point $\tilde{\mu}^{(i)}$. Analogous to the Euclidean scenario $\mathbb{R}^n$, we might know $(\eta^{(1)})^-$ (and therefore $(\eta^{(1)})^+$ by the C^1 differentiability condition settled on the first step) and wish to define iteratively $(\eta^{(i)})^-$ for $i = 2, \ldots, N-1$ (and obviously $(\eta^{(i)})^+$ in much the same way as $(\eta^{(1)})^+$).

Theorem 6.8 Consider $\tilde{\mu}^{(0)}, \ldots \tilde{\mu}^{(N)}$ as the new interpolation points on $\mathcal{P}_+(I)$, obtained from Eq. 6.96 at times $t_i = i$. Let α represent the geodesic joining two points on $\mathcal{P}_+(I)$, as given by Eq. 6.40. For a given $B^i = [((b^{(1)})^i)^-, \ldots, ((b^{(N-1)})^i)^-]^T$,

$i = 1, \ldots, N - 1$, a matrix of size $(n(N - 1) \times n)$ containing the $(N - 1)$ control points that generate the C^2 Bézier curve β_i in each tangent space $T_{\mu^{(i)}} \mathcal{P}_+(I)$, for $i = 1, \ldots, N - 1$, the Bézier spline $\gamma : [0, 1] \to \mathcal{P}_+(I)$ is of class C^2. It is uniquely defined by a set of control points given by the row of the matrix $\eta = [(\eta^{(1)})^-, \ldots, (\eta^{(N-1)})^-]^T \in \mathbb{R}^{n(N-1) \times n}$, where

1. $(\eta^{(1)})^- = Exp_{\tilde{\mu}^{(1)}}(((b^{(1)})^1)^-)$,
2. $(\eta^{(2)})^- = Exp_{(\eta^{(1)})^+} \left(\frac{1}{3} \left((d\varphi_{\tilde{\mu}^{(1)}})_{(\eta^{(1)})^-} \left(\dot{\alpha}(1, \tilde{\mu}^{(0)}, (\eta^{(1)})^-) \right) \right) - 4\dot{\alpha}(0, (\eta^{(1)})^-, \tilde{\mu}^1) \right)$,
3. $(\eta^{(i+1)})^- = Exp_{(\eta^{(i)})^+} \left(((d\varphi_{\tilde{\mu}^{(i)}})_{(\eta^{(i)})^-} \left(\dot{\alpha}(1, (\eta^{(i-1)})^+, (\eta^{(i)})^-) \right) - 2\dot{\alpha}(0, (\eta^{(i)})^-, \tilde{\mu}^{(i)})) \right)$, $i = 2, \ldots, N - 2$.

Proof The proof closely resembles to the proof of Theorem 3.3. □

6.4 Experiments

In many applications, it is of great interest to study the changes in densities as a function of time or space positions. In the following examples, we will show how the proposed methods can be used to capture dynamics as a smooth path interpolating the key observations. For example, it is essential to study the changes in densities of Regions Of Interest (ROIs) in the brain. Broad aims in the study of brain dynamics are to investigate how densities as features fold into a 3D functional path and to estimate the full range of such functional path for given functional ROIs. We illustrate this idea in Fig. 6.1 where we display two different examples. For each examples, a ROI $P_{i \leq 5}$ has a different color and is represented by a PDF. We use the methods described in Sect. 6.3 to construct C^1 and C^2 spline that interpolates $P_{i \leq 5}$ and we display the results using 25 equally spaced frames.

6.4.1 Numerical Examples

Before we show results on Brain, we illustrate C_1 and C_2 paths on PDFs manifold with several examples where densities vary from simple to complicated shapes. For this application a temporal subsequence of 5 densities equally spaced in time is taken and we consider the introduced method in Sect. 6.3 for predicting densities in between the observed ones. We display the C_1 and C_2 paths in Figs. 6.2 and 6.3.

To better visualize the differences between C_1 and C_2 solutions we first display each path as a surface where the domain of definition and time interval plays the role of x and y axes while z gives the value of the PDF. We show the surfaces from two different point of view for a better illustration. The color bar displays the amplitude of the surface. We then compute and display the norm of the derivative with respect to time.

110   6 Spline Interpolation on the Manifold of Probability Measures

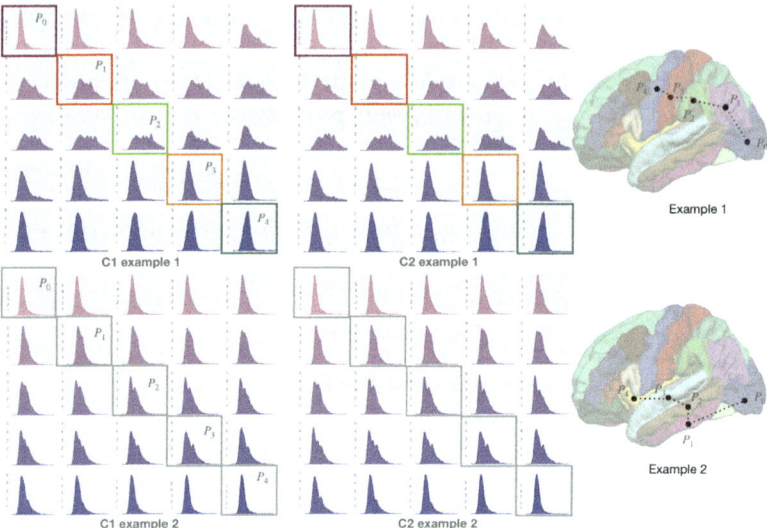

Fig. 6.1 Two illustrating examples of splines interpolating different Region of interests on the Brain

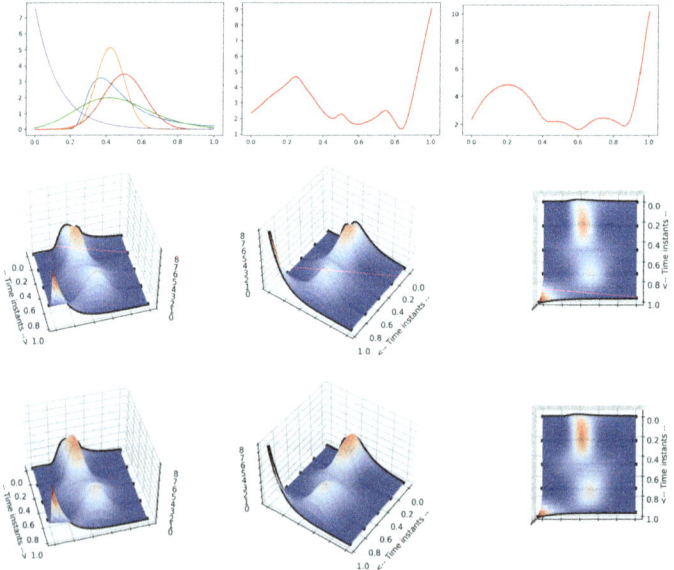

Fig. 6.2 Example 1: The first row, from left to right, displays the control PDFs, the velocity along the C^1 path, and the velocity along the C^2 path. The second row displays the C^1 path and last row displays the C^2 path as surfaces

6.4 Experiments

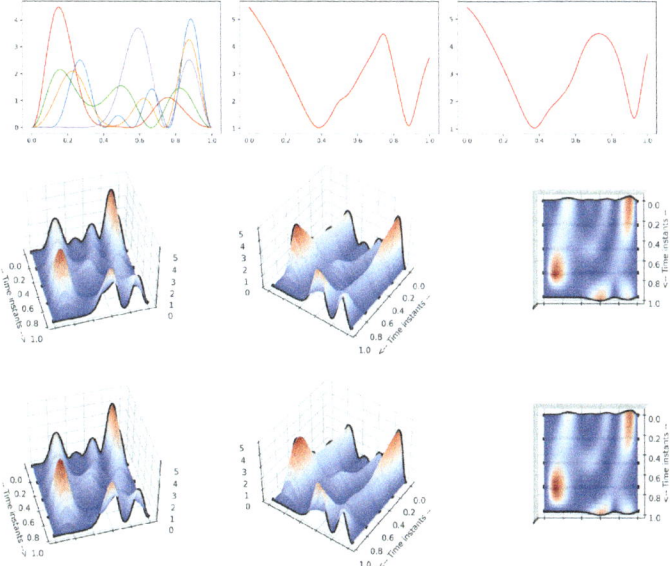

Fig. 6.3 Example 2: The first row, from left to right, displays the control PDFs, the velocity along the C^1 path, and the velocity along the C^2 path. The second row displays the C^1 path and last row displays the C^2 path as surfaces

6.4.2 Medical Examples

An important application of the proposed method is that smooth solutions between different key observations can be explored, visualized and analyzed. This is an important information about the evolution (variation) during a trajectory. To show how this could applied in a medical context we consider a dataset of averages of morphological features sampled from brain cortical surfaces for a population. The morphological measures from surfaces are derived from atlas-based registrations of individual cortices from subjects. Following registration, the cortex is parcellated into distinct regions of interest, and measures such as cortical thickness, gray matter volume, or sulcal depths over the ROIs. In this work we focus on thickness only and show trajectories crossing of the most relevant and neighboring ROIs.

An attractive feature of the spline fit is that smooth paths between different states can be explored, to investigate how possible transitions in shape from one step to another. For the smooth prediction we have used the observations at integer times but have predicted at equally spaced time points between observations. In Fig. 6.4 we display the predicted shape change in the transition to a state in a later part of the simulation using the cubic spline, at times $t = 1$ to $t = 25$ at equally spaced intervals. We can see that the smooth path predicts the density change between data points well, and that the evolution is seen in the smoothed predicted path in $\mathcal{P}$.

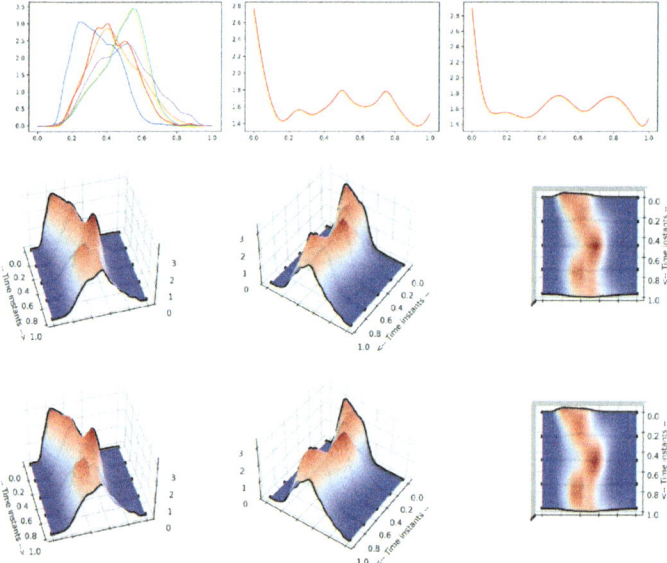

Fig. 6.4 Spline paths on Brain's ROIs: The first row from left to right displays the control PDFs, the velocity along the C^1 path, and the velocity along the C^2 path. The second row displays the C^1 path and last row the C^2 path as surfaces

References

1. S. Amari, and H. Nagaoka, Methods of Information Geometry, Translations of Mathematical Monographs, 191, 2000.
2. L. Devroye, and L. Györfi, Nonparametric Density Estimation: The L_1 View, John Wiley Sons, 1985.
3. S. T. Rachev, and L. B. Klebanov, and S. Stoyanov, and F. J. Fabozzi, The Methods of Distances in the Theory of Probability and Statistics, Springer New York, 2013.
4. C. Villani: Optimal Transport, Old and New, Springer Science & Business Media, 2009.
5. N. Ay, and J. Jost, and H. Le, and L. Schwachhofer: Information geometry, Springer, 2017.
6. Jérémie, B. and Raúl, G. and Thierry, K. and Alfredo, L.: Geodesic PCA in the Wasserstein space by convex PCA, Annales de l'Institut Henri Poincaré, 53 (1), 1–26, 2017.
7. Karimi, A. and Ripani, L. and Georgiou, T. T.: Statistical learning in wasserstein space, IEEE Control Systems Letters, 5(3), 899–904, 2021.
8. Freifel, O. and Hauberg, S. and J. Black, M.: Model Transport: Towards Scalable Transfer Learning on Manifolds, IEEE Conference on Computer Vision and Pattern Recognition, 2014.
9. Jonathan, W. and Francis, B.: Sharp asymptotic and finite-sample rates of convergence of empirical measures in Wasserstein distance, Bernoulli, 25(4A), 2620–2648, 2019.
10. Julio, B.V and Joaquin, F. and Gonzalo, R. and Felipe, T.: Bayesian learning with Wasserstein barycenters, arXiv e-prints, 2018.
11. Sinho, C. and Tyler, M. and Philippe, R. and Austin, J. S.: Gradient descent algorithms for Bures-Wasserstein barycenters, In Jacob Abernethy and Shivani Agarwal, editors, Proceedings of Thirty Third Conference on Learning Theory, 125, 1276–1304, 2020.
12. Benamou, J.D and Gallouët, T.O and Vialard, F.X: Second-Order Models for Optimal Transport and Cubic Splines on the Wasserstein Space, Foundations of Computational Mathematics volume, 19, 1113–1143, 2019.

References

13. Chen, Y. and Conforti, G. and Georgiou, T.T: Measure-Valued Spline Curves: An Optimal Transport Viewpoint, SIAM Journal on Mathematical Analysis, 50, 2018.
14. Chewi, S. and Clancy, J. and Le Gouic, T. and Rigollet, P. and Stepaniants, G. and Stromme, A.: Fast and Smooth Interpolation on Wasserstein Space, Proceedings of Machine Learning Research, 130, 2021.
15. Rao, C.R: Information and accuracy attainable in the estimation of statistical parameters, Bull. Calcutta Math. Soc, 37, 81–89, 1945.
16. Ay, N.and Jost, J. and Le, H. and Schwachhofer, L.: Information geometry, Springer, 2017.
17. Nava-Yazdani, E. and Polthier, K.:De Casteljau's algorithm on manifolds, Computer Aided Geometric Design, 30(7), 722–732, 2013.

Chapter 7
Spline Interpolation on the Manifold of Probability Density Functions

This chapter delves into the challenge of spline interpolations on $\mathcal{P}$, the space of probability density functions, especially when dealing with a limited set of observations $p_i \in \mathcal{P}$. Fitting a set of PDFs points constitutes a vital area of research in theoretical and computational statistics, with widespread applications in fields such as machine learning, medical imaging, computer vision, signal/video processing, and beyond [1–3]. Complex geometric data, originating from experiments, measurements, test methods, surveys, and other sources, are often represented as PDFs. This representation simplifies problem formulation by categorizing data based on probabilities, facilitating analysis, transmission, prediction, and classification. PDFs offer advantages such as improved visualization of local distribution, invariance to translation, uniform scaling, and rotation, and enhanced exploration of skewness in high-dimensional datasets featuring repetitive features. For instance, in scenarios involving videos or multiple images capturing the same scene at different times, PDFs can model motion, transforming the application into a regression problem on the PDF space.

The problem at hand involves a finite set of $(N + 1)$ distinct time instants t_i and corresponding data points $p_i \in \mathcal{P}$. The objective is to estimate a spline, represented as a specialized regularized function γ on $\mathcal{P}$ with $\gamma(t_i) = p_i$. The focus extends beyond generalizing splines on $\mathcal{P}$ with minimal squared-norm acceleration. We provide numerical schemes to solve for C^1 and C^2 splines from data points $p_i \in \mathcal{P}$. These methods are computationally efficient, easy to implement, extensible, and adaptable to diverse spaces and applications.

7.1 Problem Formulation

Given a finite set of increasing time instants $t_0, t_2, \ldots, t_N \in \mathbb{R}$ and Probability Density Functions (PDFs) $p_0, p_2, \ldots, p_N \in \mathcal{P}$, this chapter focuses on finding a C^2 spline γ that takes values in $\mathcal{P}$ and is a solution to the following problem:

$$\gamma : [t_0, t_N] \to \mathcal{P}, \quad \gamma(t_i) = p_i. \tag{7.1}$$

If p_i were vectors with $\|.\|_1$ or $\|.\|_2$ norms, γ could be approached as a parametric or non-parametric solution. However, in the current scenario, where p_i represents a special subset of functions, choosing the norm is non-trivial due to the structure of $\mathcal{P}$. This complexity makes the generalization of previous solutions challenging. To address this problem, we formulate γ as a spline taking values in $\mathcal{P}$, ensuring it passes through p_i at time instant t_i under differentiability constraints.

While various statistical models, such as maximum likelihood and Bayes estimators, have been introduced in the literature for learning and predicting Probability Density Functions (PDFs), nonparametric fitting remains a challenging problem, particularly in terms of computational efficiency. Meanwhile, radial basis functions are widely employed for interpolations [4]. PDFs inherently possess infinite-dimensional characteristics, representing a Riemannian manifold [5] structure. Consequently, extending traditional regression methods from finite vectors to functional instances is not a straightforward task [6]. This challenge has prompted the development of diverse frameworks for comparing PDFs, employing various metrics such as Frobenius, Fisher–Rao, log-Euclidean, Jensen–Shannon, and Wasserstein metrics [7]. The primary objective of this chapter is to introduce a novel framework for fitting a given set of nonparametric PDFs within the context of spline regression. The innovation in this chapter and the proposed methods lies in leveraging the Riemannian structure of the space of PDFs, significantly enhancing the computational tractability of the solution.

More precisely, let $p_0, \ldots, p_N$ be an indexed set of probability density functions associated to a set of observation times $t_0, \ldots, t_N$ such that $t_i = i$, for $i = 0, \ldots, N$. The goal is to estimate a path $\gamma : [0, N] \to \mathcal{P}$ that minimizes the following functional

$$E(\gamma) = \frac{\lambda}{2} \int_{t_0}^{t_N} < \frac{D^2\gamma(t)}{Dt^2}, \frac{D^2\gamma(t)}{Dt^2} >_\mathcal{P} + \frac{1}{2} \sum_{i=0}^{N} d_\mathcal{P}^2(\gamma(t_i), p_i), \tag{7.2}$$

where $\lambda (> 0)$ is a smoothing parameter, $< ., . >_\mathcal{P}$ and $d_\mathcal{P}$ denote the Fisher–Rao metric and the geodesic distance on $\mathcal{P}$. In this chapter, the goal is to find a solution γ to the optimization problem (7.2), structured as a Bézier spline, that satisfies two key properties: (i) $\gamma(t_i) = p_i$ and (ii) γ belongs to the class C^2. Working directly with $\mathcal{P}$ equipped with the Fisher–Rao metric can be computationally challenging, primarily due to the intricate nature of computing geodesics between PDFs on $\mathcal{P}$. To overcome this challenge, we shift the focus to the square-root density representation $\psi_i = \sqrt{p_i}$ for each PDF p_i, where $\mathcal{M}$ denotes the space containing all ψ_is. Notably, this resulting space is a subset of the unit sphere in a Hilbert space, equipped with the standard L^2 metric. By opting for the square-root representation ψ_i of p_i, we first address the challenge by finding a C^2 Bézier spline on the symmetric space $\mathcal{M}$. The computational advantage of working in this space lies in its ease of computing geodesics, Riemannian exponential and logarithmic maps, and parallel transports. Once this task is accomplished, we utilize the points along the spline on $\mathcal{M}$ to determine an interpolating spline γ on $\mathcal{P}$. Importantly, we establish that $\mathcal{P}$ is a locally symmetric space, ensuring C^2 continuity and optimizing the resulting solution on $\mathcal{P}$.

In this chapter, we initiate the discussion from the energy minimization formulation of linear least-squares in Euclidean spaces and extend this concept to the space of Probability Density Functions (PDFs). As far as we know, this framework represents the first attempt to address the fitting problem in the context of splines within the space of PDFs. We demonstrate that the proposed method offers geometric simplicity, extensibility, and computational efficiency.

7.2 Geometry of the Space of Probability Density Functions

In this section, we revisit the Riemannian structure inherent in the space of Probability Density Functions (PDFs). These foundational concepts are extensively discussed in existing literature, particularly in [7, 8]. Subsequently, we adapt insights from [9] to derive pertinent geometric tools, including geodesics, the Riemannian Exponential (Exp) and Logarithmic (Log) maps. Notably, the focus will be on the square-root representation of PDFs. It is demonstrated that this representation proves to be efficient, where the Fisher–Rao metric simplifies to a straightforward L^2 metric, and the space of constrained non-negative continuous functions emerges as a submanifold of the unit Hilbert sphere S^∞.

7.2.1 Fisher–Rao Geometry

Let $I = [0, 1]$ be the unit interval, and let $p : I \to \mathbb{R}$ denote a continuous probability density function over I. Therefore, the space of probability density functions is defined as

$$\mathcal{P} = \{p : I \to \mathbb{R} \mid \forall t, \ p(t) \geq 0, \int_0^1 p(t)dt = 1\}. \tag{7.3}$$

It is important to note that we refrain from imposing any significant assumptions on the functional form of the density, denoted as $p \in \mathcal{P}$. Consequently, $\mathcal{P}$ encompasses nonparametric models. Additionally, it has been demonstrated that $\mathcal{P}$ is an infinite-dimensional smooth manifold with a boundary; further details can be found in [9],

Sect. 4.11.2, p. 113. To elaborate, considering a density $p \in \mathcal{P}$, the tangent space is determined by differentiating the unit condition on $\mathcal{P}$, resulting in:

$$T_p\mathcal{P} = \{f \in L^1(I, \mathbb{R}) \mid \int_0^1 f(t)dt = 0\}. \tag{7.4}$$

We may now endow $\mathcal{P}$ with a Riemannian metric. In this chapter, we are interested in the Fisher–Rao metric defined, for a $p \in \mathcal{P}_+ = \{p \in \mathcal{P} \mid p > 0\}$ and vectors for any tangent vectors $f_1, f_2 \in T_p\mathcal{P}$ by

$$\langle f_1, f_2 \rangle_p = \int_0^1 f_1(t) f_2(t) \frac{1}{p(t)} dt. \tag{7.5}$$

The geodesic distance between two nonparametric densities under the Fisher–Rao metric is given by

$$d_\mathcal{P}(p_1, p_2) = \inf_\Gamma L[\eta], \tag{7.6}$$

where Γ is the set of all geodesic paths $\eta : I \to \mathcal{P}$ joining p_1 and p_2 such that $\eta(0) = p_1$ and $\eta(1) = p_2$ and $L[\eta] = \int_0^1 \sqrt{\langle \dot{\eta}(t), \dot{\eta}(t) \rangle_p}$ is the length of η.

7.2.2 Riemannian Isometry

The Fisher–Rao metric, initially introduced in [10], was later proven to be the unique intrinsic metric on the probability density functions (PDFs) manifold $\mathcal{P}$ in [11]. This exploration of the geometry of the infinite-dimensional manifold, equipped with the Riemannian structure, is commonly referred to as information geometry. Recent research has underscored the significance and utility of the geometric structure of the smooth manifold $\mathcal{P}$ endowed with the Fisher–Rao metric in various applications within statistics and information theory. Many of these applications leverage crucial geometric tools like the Riemannian exponential map, its inverse, means, geodesic arcs, among others.

Due to the nonlinear constraints imposed on these probability density functions, performing calculus directly on the representation of the space $\mathcal{P}$ can be highly challenging. Consequently, considerable efforts have been directed toward developing alternative representations of the space $\mathcal{P}$ using certain admissible functions. Examples include distributions, log density functions, and square-root density functions. Notably, the Fisher–Rao metric remains invariant across all parametrizations of the PDFs space. Among these representations, the manifold comprised of square-root density functions has been identified as numerically the most suitable for practical applications.

For a given $p \in \mathcal{P}$, the square-root density function is defined as $\psi = \sqrt{p}$. When restricted to $\psi \geq 0$, and considering that $\int_0^1 \psi^2(t)dt = \int_0^1 p(t)dt = 1$, it becomes

7.2 Geometry of the Space of Probability Density Functions

evident that ψ belongs to the closed spherically convex set $\mathcal{S}_+^\infty$ within the infinite-dimensional unit sphere $\mathcal{S}^\infty$. More precisely, the space of all square-root density functions, denoted as $\mathcal{M}$, is expressed as:

$$\mathcal{M} = \{\psi \in L^2(I, \mathbb{R}) \mid \psi \geq 0, \int_0^1 \psi^2(t)dt = 1\}. \tag{7.7}$$

We check at once that the tangent space $T_\psi \mathcal{M}$ at each $\psi \in \mathcal{M}$ can be identified with, (see [9], Sect. 4.3, p. 82),

$$T_\psi \mathcal{M} = \{v \in L^2(I, \mathbb{R}) \mid \int_0^1 v(t)\psi(t)dt = 0\}. \tag{7.8}$$

Remarkably, using the Riemannian manifold structure for the square root of a probability density function (PDF) allows for the direct resolution of crucial problems. These include regression, clustering, classification, fitting, and interpolation of data on $\mathcal{P}$. In this simplified space $\mathcal{M}$, which is isometrically equivalent to $\mathcal{P}$ through a transformation, well-established Riemannian tools such as the geodesic, exponential map, log map, and parallel transport are readily applicable.

Lemma 7.1 *The map $\Phi : \mathcal{P} \to \mathcal{M}$ defined by $\Phi(p) := 2\sqrt{p}$ is a Riemannian isometry.*

Proof It is clear that Φ is well-defined since p is nonnegative and Φ is a diffeomorphism from the interior of $\mathcal{P}$ to $\mathcal{M}$ without the boundary. Moreover, $\Phi^{-1}(\psi) = \frac{1}{4}\psi^2$. To check that Φ is an isometry, let $p \in \mathcal{P}$. Compute the differential of Φ at p, $d\Phi_p : T_p\mathcal{P} \to T_{\Phi(p)}\mathcal{M}$,

$$(d\Phi_p)(f) = \frac{d\Phi_p}{dp}(f) = \frac{f}{\sqrt{p}}. \tag{7.9}$$

Hence, it suffices to prove that for any $f_1, f_2 \in T_p\mathcal{P}$, $\langle d\Phi_p(f_1), d\Phi_p(f_2)\rangle_\mathcal{M} = \langle f_1, f_2\rangle_\mathcal{P}$.

An easy computation gives,

$$\langle d\Phi_p(f_1), d\Phi_p(f_2)\rangle_\mathcal{M} = \int_0^1 d\Phi_p(f_1(t))d\Phi_p(f_2(t))dt$$
$$= \int_0^1 \frac{f_1(t)}{\sqrt{p(t)}} \frac{f_2(t)}{\sqrt{p(t)}}dt$$
$$= \int_0^1 \frac{f_1(t)f_2(t)}{p(t)}dt$$
$$= \langle f_1, f_2\rangle_\mathcal{P},$$

and the lemma follows. □

7.2.3 Riemannian Structure of $\mathcal{M}$

Here we briefly recall the differential structure of the Riemannian manifold $\mathcal{M}$ and set up without proof some geometric structures that will be used in the next section.

Given a point $\psi_1 \in \mathcal{M}$ and a tangent vector $v \in T_\psi \mathcal{M}$, there exists a unique geodesic $t \to \alpha_t(\psi_1; v)$ starting from a point ψ_1 with direction v

$$\alpha_t(\psi_1; v) = \cos(t\|v\|_2)\psi_1 + \sin(t\|v\|_2)\frac{v}{\|v\|_2}, \quad t \in [0, 1], \tag{7.10}$$

where $\|.\|_2$ denotes the L^2 norm. We write

$$\dot{\alpha}_t(\psi_1; v) = \frac{\partial}{\partial u}|_{u=t}\alpha_t(\psi_1; v). \tag{7.11}$$

Moreover, the geodesic arc joining two points (not antipodal) ψ_1 (at $t=0$) and ψ_2 (at $t=1$) is given by:

$$\alpha_t(\psi_1; \psi_2) = \frac{\sin\left((1-t)\cos^{-1}\langle\psi_1, \psi_2\rangle_\mathcal{M}\right)}{\sin\left(\cos^{-1}\langle\psi_1, \psi_2\rangle_\mathcal{M}\right)}\psi_1 + \frac{\sin\left(t\cos^{-1}\langle\psi_1, \psi_2\rangle_\mathcal{M}\right)}{\sin\left(\cos^{-1}\langle\psi_1, \psi_2\rangle_\mathcal{M}\right)}\psi_2. \tag{7.12}$$

Consequently, the exponential map $\mathrm{Exp}_{\psi_1}: T_\psi \mathcal{M} \to \mathcal{M}$ is defined as

$$\mathrm{Exp}_{\psi_1}(v) = \psi_2 = \cos(\|v\|_2)\psi_1 + \sin(\|v\|_2)\frac{v}{\|v\|_2}. \tag{7.13}$$

Henceforth, the derivative of $\mathrm{Exp}\psi 1$ at $v \in T_{\psi_1}\mathcal{M}$ will be represented as $(d\mathrm{Exp}\psi 1)v$. By enforcing the condition $\|v\|2 < \pi$, the exponential map becomes injective. Likewise, for two points ψ_1 and ψ_2 on $\mathcal{M}$, the inverse exponential map (also referred to as the logarithmic map) is computed as follows:

$$\mathrm{Exp}_{\psi_1}^{-1}(\psi_2) = f = \frac{\cos^{-1}\langle\psi_1, \psi_2\rangle_\mathcal{M}}{\sin(\cos^{-1}\langle\psi_1, \psi_2\rangle_\mathcal{M})}\left(\psi_2 - \langle\psi_1, \psi_2\rangle_\mathcal{M}\psi_1\right). \tag{7.14}$$

The parallel transport of a vector f along a geodesic curve $\alpha_t(\psi_1, f)$ on $\mathcal{M}$ joining $\alpha(0) = \psi_1$ and $\alpha(1) = \psi_2$ is the linear isometry $\Gamma: T_{\psi_1}\mathcal{M} \to T_{\psi_2}\mathcal{M}$ given by

$$\Gamma_{\psi_1 \mapsto \psi_1}(f) = f - 2\frac{(\psi_1 + \psi_2)}{\|\psi_1 + \psi_2\|_2^2}\langle f, \psi_2\rangle_\mathcal{M}. \tag{7.15}$$

The symmetry of the Riemannian manifold $\mathcal{M}$ is another noteworthy property essential for proving the main results of this chapter in the subsequent section. Specifically, for any $\psi_1 \in \mathcal{M}$, the geodesic symmetry that reverses geodesics passing through ψ_1 is expressed as:

$$\varphi_{\psi_1}(\psi_2) = -\psi_2 + 2\langle\psi_2, \psi_1\rangle_\mathcal{M}\psi_1, \tag{7.16}$$

7.2 Geometry of the Space of Probability Density Functions

which is also an isometry of $\mathcal{M}$. In what follows, we denote by $(d\varphi_{\psi_1})_{\psi_2}$ the derivative of φ_{ψ_1} at ψ_2.

7.2.4 Geodesic Curves on $\mathcal{P}$

We introduce several foundational lemmas to demonstrate the transference of the L^2 geometry from $\mathcal{M}$ to the space of nonparametric densities $\mathcal{P}$. Initially, we note that the Fisher–Rao metric on $\mathcal{P}$ transforms into the conventional L^2 metric on $\mathcal{M}$, up to a constant. More precisely, for any $v_1, v_2 \in T_\psi \mathcal{M}$, we find:

$$\begin{aligned}
\langle f_1, f_2 \rangle_\mathcal{P} &= \int_0^1 f_1(t) f_2(t) \frac{1}{p(t)} dt \\
&= \int_0^1 2\sqrt{p(t)} v_1(t) 2\sqrt{p(t)} v_2(t) \frac{1}{p(t)} dt \\
&= 4 \int_0^1 v_1(t) v_2(t) dt \\
&= <\tilde{v}_1, \tilde{v}_1>_\mathcal{M}.
\end{aligned}$$

Moreover, the previously numerically challenging geodesic distance $d_\mathcal{P}(p_1, p_2)$ between two PDFs, calculated under the Fisher–Rao metric, is now straightforwardly expressed through the intrinsic geodesic distance $d_\mathcal{M}(\psi_1, \psi_2)$ on the unit Hilbert sphere between two distinct and non-antipodal points $\psi_1, \psi_2 \in \mathcal{M}$, as follows:

$$d_\mathcal{P}(p_1, p_2) = \cos^{-1}(<\psi_1, \psi_2>_\mathcal{M}). \tag{7.17}$$

Proposition 7.1 *Let $\Phi : \mathcal{P} \to \mathcal{M}$ be the Riemannian isometry. Let Φ^* denote the differential of Φ at a point $p \in \mathcal{P}$.*

1. *Φ takes geodesics to geodesics: if $t \to \eta_t(p_1; f)$ is the geodesic on $\mathcal{P}$ starting from a point p_1 with direction f, then $\Phi \circ \eta$ is the geodesic on $\mathcal{P}$ with initial point ψ and initial direction $\Phi^* f$.*
2. *If $Exp_p : T_\psi \mathcal{P} \to \mathcal{P}$ is the exponential map defined at a point $p \in \mathcal{P}$ and $Exp_\psi : T_\psi \mathcal{M} \to \mathcal{M}$ is the exponential map defined at a point $\Phi(p) = \psi \in \mathcal{M}$, then we have:*
$$\Phi = Exp_p \circ \Phi^* \circ Exp_\psi \tag{7.18}$$

The significance of the previous proposition lies in its capacity to reformulate the geodesic Eq. 7.10 on $\mathcal{M}$ in terms of square roots, thereby deriving the corresponding equation on $\mathcal{P}$ (see Fig. 7.2).

Lemma 7.2 *The geodesic $t \to \eta_t(p_1; f)$ starting from a point $p \in \mathcal{P}$ with direction $f \in T_p\mathcal{P}$ takes the form*

$$\eta_t(p; f) = \left[\cos\left(t\sqrt{<f,f>_p}\right) \sqrt{p} + \sin\left(t\sqrt{<f,f>_p}\right) \frac{f}{\sqrt{<f,f>_p}} \right]^2. \tag{7.19}$$

Proposition 7.2 *Let $t \to \gamma(t)$ be a curve on $\mathcal{P}$ and $f \in T_p\mathcal{P}$ a vector field along γ. Let $\Phi : \mathcal{P} \to \mathcal{M}$ be the Riemannian isometry, then*

1. *The covariant derivative of a vector fields along γ is preserved, that is*

$$\Phi^*(D_t f) = D_t(\Phi^* f). \tag{7.20}$$

2. *If Υ_γ denotes the parallel transport along the curve γ and $\Upsilon_{\Phi \circ \gamma}$ denotes the parallel transport along the curve $\Phi \circ \gamma$ then,*

$$d\Phi_{\gamma(1)} \circ \Upsilon_\gamma = \Upsilon_{\Phi \circ \gamma} \circ d\Phi_{\gamma(0)}. \tag{7.21}$$

3. *The Riemannian curvature endomorphism is isometry invariant, that is*

$$d\Phi_p R(f_1, f_2, f_3) = R(d\Phi_p f_1, d\Phi_p f_2) d\Phi_p f_3, \quad \forall f_i \in T_p\mathcal{P}, i = 1, 2, 3. \tag{7.22}$$

This result will be extremely useful to prove the next Theorem.

Theorem 7.1 *$\mathcal{P}$ equipped with the Fisher–Rao metric is a locally symmetric space.*

To show that $\mathcal{P}$ is a locally symmetric space, we need the following proposition.

Proof (Proof of Theorem 7.1) It follows by Lemma 7.1 as $\mathcal{P}$ is locally isometric to the symmetric space $\mathcal{M}$. Hence, around any point $p \in \mathcal{P}$ there is a geodesic ball $B_r(p)$ which is isometric to a ball $B_\epsilon(\psi)$ on $\mathcal{M}$. Therefore, there is an isometry $\kappa_p : B_r(p) \to B_r(p)$ such that $\kappa_p(p) = p$ and the differential $d\kappa_p(p) = -Id$. Let $f_1, f_2, f_3, f_4 \in T_p(\mathcal{P})$. Let us show that $\nabla_{f_1} R(f_2, f_3, f_4) = 0$. Set $f = \nabla_{f_1} R(f_2, f_3, f_4)$ and applying $d\kappa_p$ on both sides of this equation, we obtain that f will be changed by $-f$ and f_i to $-f_i$, for $i = 1, \ldots, 4$. Thus the left-hand side changes sign while the right-hand side stays the same. Moreover, we mention that the curvature tensor is preserved by $d\kappa_p$. Consequently, $\nabla R = 0$ and the proof is complete. □

In the subsequent section, we will demonstrate how to leverage these findings to formulate an intrinsic geometric algorithm to construct a C^2 Bézier spline on the space of probability density functions $\mathcal{P}$.

7.3 Interpolation Problem on the Space of Probability Density Functions

Consider an indexed set of probability density functions $p_0, \ldots, p_N$ associated with a corresponding set of observation times $t_0, \ldots, t_N$, where $t_i = i$ for $i = 0, \ldots, N$. The primary objective is to identify a C^2 spline $\gamma : [t_0, t_N] \to \mathcal{P}$ that interpolates the provided set of probability density functions, i.e., $\gamma(t_i) = p_i$ for $i = 0, \ldots, N$. Due to the numerical challenges posed by $\mathcal{P}$ equipped with the Fisher–Rao metric, we adopt an indirect approach. Instead of directly seeking a solution for the cost function (7.2) on $\mathcal{P}$, we reformulate the problem and address it on the space of all square-root density functions, denoted as $\mathcal{M}$. In essence, the approach unfolds as follows: By computing $\psi_0, \ldots, \psi_N$, the square-root density representations of each p_i for $i = 1, \ldots, N$, the goal is to find a C^2 interpolating Bézier spline $\sigma : [t_0, t_N] \to \mathcal{M}$ that minimizes the following cost function:

$$\min_{\sigma \in C^2([t_0, t_N], \mathcal{M})} E(\sigma) = \int_{t_0}^{t_N} < \frac{D^2\sigma(t)}{Dt^2}, \frac{D^2\sigma(t)}{Dt^2} >_{\mathcal{M}} . \quad (7.23)$$

To achieve this objective, we initiate the process by formulating a C^1 Bézier interpolation spline on $\mathcal{M}$. Subsequently, taking into consideration the global symmetries present at interpolation points, we derive equations for the control points that generate the desired C^2 Bézier spline. Once this step is completed, we proceed to compute the spline σ on $\mathcal{M}$ in order to obtain an interpolating spline γ on $\mathcal{P}$. Additionally, we demonstrate that $\mathcal{P}$ is a locally symmetric space. Consequently, the C^2 continuity is guaranteed at the joint point p_i, and the resulting interpolating spline on $\mathcal{P}$ is considered optimal.

7.3.1 C^1 Spline on $\mathcal{M}$

Let us consider $\sigma_j : [t_0, t_N] \to \mathcal{M}$ the Bézier curve of order j defined by $(j + 1)$ control points $V_0, \ldots, V_j \in \mathcal{M}$. Consider the point $V_i^0 = V_i$, and for $i = 0, \ldots, j - k, k = 1, \ldots, j$ we define

$$V_i^k = \sigma_k(t, V_i, \ldots, V_{i+k}) = \alpha\left(t, V_i^{k-1}, V_{i+1}^{k-1}\right), t \in [0, 1],$$

the ith point of the kth step of the De Casteljau algorithm, and thus $\alpha_j(t; V_0, \ldots, V_j) = V_0^j$. Here, α represents the geodesic curve on $\mathcal{M}$ given by Eq. 7.10. In this context, the Bézier spline $\sigma : [t_0, t_N] \to \mathcal{M}$ is constructed as a sequence of n Bézier curves σ_j^i of orders two and three. The first and the last curves are quadratic Bézier curves, while all the others are cubic. Additionally, let $(\widehat{\chi}_i^-, \widehat{\chi}_i^+)$

denote the control points on the left and right-hand sides of the interpolation point ψ_i for $i = 1, \ldots, (N-1)$. The Bézier spline $\sigma : [t_0, t_N] \longrightarrow \mathcal{M}$ is then given by

$$\sigma(t) = \begin{cases} \sigma_2^0(t; \psi_0, \widehat{\chi}_1^-, \psi_1), & 0 \le t \le 1, \\ \sigma_3^i(t-i; \psi_i, \widehat{\chi}_i^+, \widehat{\chi}_{i+1}^-, \psi_{i+1}), & i-1 \le t \le i \\ \sigma_2^{N-1}(t-(N-1); \psi_{N-1}, \widehat{\chi}_{N-1}^+, \psi_N), & N-1 \le t \le N \end{cases}$$

Since the Bézier spline σ interpolates the first and the last control points of each Bézier curve σ_j^i, $j \in \{2, 3\}$, $0 \le i \le N-1$, the continuity of σ at joint points is well satisfied. However, as mentioned in the last section, we ideally want σ to be of class C^2. To achieve this, the algorithm proceeds in two main steps: first, solve a C^1 Bézier spline on $\mathcal{M}$, and then use this solution to construct a C^2 spline.

7.3.2 C^2 Bézier Splines on $\mathcal{M}$

Let us first establish the conditions required to ensure C^1 continuity along the curve. To address this concern, we propose to shift the problem to the tangent space $T_{\psi_i}\mathcal{M}$, for $i = 1, \ldots, N-1$, and then bring back the solution to the Riemannian manifold $\mathcal{M}$. The proposed algorithm relies solely on the Riemannian exponential and logarithm maps. Specifically, given $\psi_0, \ldots, \psi_N$ as an indexed set of square-root density functions on $\mathcal{M}$, we use the Riemannian logarithmic map, as defined by Eq. 7.14, to lift data points $\psi_0, \ldots, \psi_N$ to each tangent space $T_{\psi_i}\mathcal{M}$, $i = 1, \ldots, N-1$. The mapped data are then defined by $\phi^i = (\phi_0^i, \ldots, \phi_N^i)$ with $\phi_k^i = \text{Log}_{\psi_i}(\psi_k)$ for $k = 0, \ldots, n$. Let $\beta : [t_0, t_N] \to T_{\psi_i}\mathcal{M}$ denote the Bézier spline on $T_{\psi_i}\mathcal{M}$, $i = 1, \ldots, N-1$, defined by N Bézier curves β_j^i, $j \in \{2, 3\}$, $0 \le i \le N-1$. Hence, the optimization problem (7.23) can be formulated as follows on $T_{\psi_i}\mathcal{M}$:

$$\min_{(b_1^i)^-, \ldots, (b_{N-1}^i)^-} E((b_1^i)-, \ldots, (b_{N-1}^i)^-) := \min_{(b_1^i)^-, \ldots, (b_{N-1}^i)^-} \int_0^1 \|(\ddot{\beta}_2^i)^0(t; \phi_0^i, (b_1^i)^-, \phi_1^i)\|_2^2 +$$
$$\sum_{i=1}^{N-2} \int_0^1 \|\ddot{\beta}_3^i(t; \phi_i^i, (b_i^i)^+, (b_{i+1}^i)^-, \phi_{i+1}^i)\|_2^2 + \int_0^1 \|\ddot{\beta}_2^{N-1}(t; \phi_{N-1}^i, (b_{N-1}^i)^+, \phi_N^i)\|_2^2.$$

(7.24)

where $(b_1^i)^-$ and $(b_1^i)^+$ denote control points on the left and on the right-hand side of the interpolation point ϕ_i^i for $i = 1, \ldots, (N-1)$ on $T_{\psi_i}\mathcal{M}$. Since the tangent space is a vector space, problem (7.24) is treated similarly to the Euclidean case $\mathcal{M} = \mathbb{R}^n$. In this case, we prove that the solutions to the problem of minimizing the mean square acceleration of the Bézier curve β are exactly the control points of the curve. Additionally, we provide conditions under which the Bézier curve β is of class C^2.

7.3 Interpolation Problem on the Space of Probability Density Functions

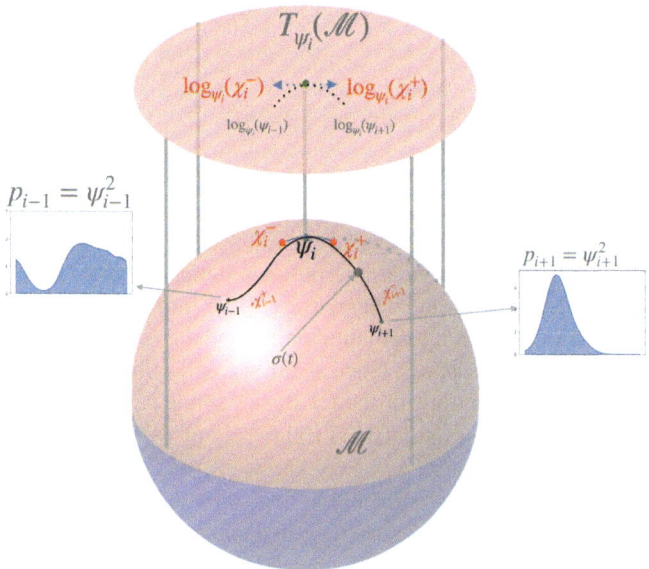

Fig. 7.1 Geometrical illustration of the Riemannian manifold $\mathcal{M}$ and its tangent space $T_{\psi_i}(\mathcal{M})$ at ψ_i. σ is an interpolating spline with $\psi_t = \sigma(t)$ for $t \in I$ which verifies $\psi_t \geq 0$, $\int_I \psi_t^2 = 1$, and $\psi_{t_i} = \psi_i$

The details to obtain equations that govern the control points of the C^2 Bézier spline on $\mathbb{R}^N$ are given in chapter 1. Finally, the Riemannian exponential map Exp_{ψ_i} defined on $\mathcal{M}$ by Eq. 7.13 will move the control points of the C^2 Bézier curve β from the tangent space $T_{\psi_i}\mathcal{M}$ to $\mathcal{M}$. Moreover, the resulting control points in $\mathcal{M}$ are optimal, and we assert that the curve is of class C^1. A geometrical illustration (sketch) is provided in Fig. 7.1.

Theorem 7.2 *The Bézier spline $\sigma : [t_0, t_N] \to \mathcal{M}$ interpolating the data points ψ_i on $\mathcal{M}$ is of class C^1 and is uniquely defined by the set of control points $\widehat{\chi} = [\widehat{\chi}_1^-, \ldots, \widehat{\chi}_{N-1}^-]^T \in \mathbb{R}^{n(N-1) \times n}$ given by:*

$$\widehat{\chi}_i^- = \text{Exp}_{\psi_i}(\tilde{x}_i), \ i = 1, \ldots, N-1, \tag{7.25}$$

where $\tilde{x}_i$, represent the row i of $B^i = [((b_1)^i)^-, \ldots, ((b_{N-1})^i)^-]^T$ in $T_{\psi_i}\mathcal{M}$, for $i = 1, \ldots, N - 1$.

Proof The proof is similar to the proof of Theorem 3.2. □

Remark 7.1 The interpolation point ϕ_N^i is adjusted to satisfy the C^2 differentiability condition of the curve β_i on $T_{\psi_i}\mathcal{M}$ for $i = 1, \ldots, N - 1$. Consequently, the point ψ_N is updated, and the new set of $(N + 1)$ interpolation points on $\mathcal{M}$ is determined as follows:

$$\tilde{\psi}_k = \mathrm{Exp}_{\tilde{\psi}_i}(\tilde{\phi}_k^i), \quad k = 0, \ldots, N; \quad i = 1, \ldots, N-1, \qquad (7.26)$$

where $\tilde{\phi} = [\tilde{\phi}_0^i, \ldots, \tilde{\phi}_N^i]^T$ is a matrix of size $n(N+1) \times n$ containing the new $(N+1)$ interpolation points in each tangent space $T_{\tilde{\psi}_i}\mathcal{M}$.

The objective now is to demonstrate that Bézier curves σ_j^i, $j \in \{2, 3\}$, $i = 0, \ldots, N-1$, can be arranged such that C^2 differentiability at joint points is satisfied, allowing σ to be a C^2 spline. The key geometric element employed to achieve this goal is the global symmetries at interpolation points provided on the Riemannian manifold $\mathcal{M}$. Lemma 4.2 and Theorem 4.1 delve into the detailed relationship between global symmetries at interpolation points and the C^2 differentiability of the Bézier spline on symmetric spaces. Once again, we will utilize this pertinent material to compute the control points of the C^2 Bézier spline σ on $\mathcal{M}$. Let χ_i^- and χ_i^+ be the new control points on the left and right sides of the interpolation point $\tilde{\psi}_i$. Similar to the Euclidean case $\mathbb{R}^n$, we may know χ_1^- (and therefore χ_1^+ by the C^1 differentiability condition established in the first step) and aim to iteratively define χ_i^- for $i = 2, \ldots, N-1$ (and obviously χ_i^+ in a manner similar to χ_1^+).

Theorem 7.3 *Consider $\tilde{\psi}_0, \ldots \tilde{\psi}_N$ the new interpolation points on $\mathcal{M}$ given by Eq. 7.26 at times $t_i = i$ and $\alpha(t)$ the geodesic joining two points on $\mathcal{M}$ given by Eq. 7.10. For a given $B^i = [(b_1^i)^-, \ldots, (b_{N-1}^i)^-]^T$, $i = 1, \ldots, N-1$, a matrix of size $(n(N-1) \times n)$ containing the $(N-1)$ control points that generate the C^2 Bézier curve β_i in each tangent space $T_{\tilde{\psi}_i}\mathcal{M}$, for $i = 1, \ldots, N-1$, the Bézier spline $\sigma : [t_0, t_N] \to \mathcal{M}$ is C^2 and is uniquely defined by a set of control points given by the row of the matrix $\chi = [\chi_1^-, \ldots, \chi_{N-1}^-]^T \in \mathbb{R}^{n(N-1) \times n}$ by*

(i) $\chi_1^- = \mathrm{Exp}_{\tilde{\psi}_1}((b_1^1)^-)$,

(ii) $\chi_2^- = \mathrm{Exp}_{\chi_1^+}\left(\frac{1}{3}\left((d\varphi_{\tilde{\psi}_1})_{\chi_1^-}\left(\dot{\alpha}(1, \tilde{\psi}_0, \chi_1^-)\right) - 4\dot{\alpha}(0, \chi_1^-, \tilde{\psi}_1)\right)\right)$,

(iii) $\chi_{i+1}^- = \mathrm{Exp}_{\chi_i^+}\left(\left((d\varphi_{\tilde{\psi}_i})_{\chi_i^-}\left(\dot{\alpha}(1, \chi_{i-1}^+, \chi_i^-)\right) - 2\dot{\alpha}(0, \chi_i^-, \tilde{\psi}_i)\right)\right)$,
$i = 2, \ldots N-2$.

Proof The proof follows a similar reasoning as the proof of Theorem 3.3. □

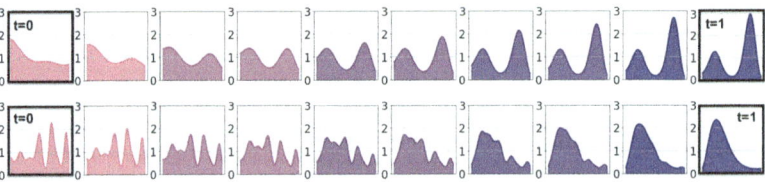

Fig. 7.2 Two examples of geodesic curves starting from a PDF at the extreme left ($t = 0$) and ending at a different PDF at the extreme right ($t = 1$). Intermediate PDFs are displayed at uniform time instants $t_i \in]0, 1[$ from the geodesic path

7.3.3 Properties of Splines on $\mathcal{P}$

We are now ready to present one of the main outcomes of this chapter, which provides the corresponding equations, akin to those derived in the previous section, that determine the control points of a C^2 Bézier spline on $\mathcal{M}$.

Theorem 7.4 *Consider $p_0, \ldots, p_N$ an indexed set of probability density functions associated with observation times $t_i = i$, for $i = 0, \ldots, N$. Let $\chi = [\chi_1^-, \ldots, \chi_{N-1}^-]^T$ be the matrix of size $n(N-1) \times n$ that generates control points of C^2 Bézier spline σ on $\mathcal{M}$. Then, the Bézier spline $\gamma : [t_0, t_N] \to \mathcal{P}$ satisfies the following properties:*

(i) $\gamma(t_i) = p_i$, $i = 0, \ldots, N$.
(ii) γ *is uniquely defined by* $(N-1)$ *control points* $Q = [q_1^-, \ldots, q_{N-1}^-]^T \in \mathbb{R}^{N(N-1) \times N}$, *where the rows of Q are given by:*

$$q_i^- = \Phi^{-1}(\chi_i^-), \ i = 1, \ldots, N-1. \tag{7.27}$$

(iii) γ *is C^2 on $\mathcal{P}$.*

Proof (i) and (ii) follows from Lemma 7.19 means that γ will be also defined by n Bézier curves γ_i^j, $j \in \{2, 3\}$, $0 \le i \le N-1$, image by the isometry Φ^{-1} of σ_i^j on $\mathcal{M}$ and control points that generate the spline γ are equivalently images of χ_i^-, namely

$$q_i^- = \Phi^{-1}(\chi_i^-), \ i = 1, \ldots, N-1. \tag{7.28}$$

The resulting Bézier spline γ is then reconstructed with De Casteljau algorithm. We should mention that the interpolation points p_i on $\mathcal{P}$ will change, since ψ_i are modified on $\mathcal{M}$. We obtain

$$\tilde{p}_i = \Phi^{-1}(\tilde{\psi}_i), \ i = 0, \ldots, N. \tag{7.29}$$

Let us prove (iii). It is clear that $\gamma|_{[t_i, t_{i+1}]}$ is C^∞. Thus, it remains to prove that γ is of class C^2. It suffices to write the C^2 differentiability condition at joint points $\tilde{p}_i$. For instance, at $\tilde{p}_1$, we need to show that

$$\frac{D}{dt}\Big|_{t=1} \dot{\gamma}_2^0(t; \tilde{p}_0, q_1^-, \tilde{p}_1) = \frac{D}{dt}\Big|_{t=0} \dot{\gamma}_3^1(t; \tilde{p}_1, q_1^+, q_2^-, \tilde{p}_2). \tag{7.30}$$

The curve σ on $\mathcal{M}$ is of class C^2, hence at $\tilde{\psi}_1$, we have

$$\frac{D}{dt}\Big|_{t=1} \dot{\sigma}_2^0(t; \tilde{\psi}_0, \chi_1^-, \tilde{\psi}_1) = \frac{D}{dt}\Big|_{t=0} \dot{\sigma}_3^1(t; \tilde{\psi}_1, \chi_1^+, \chi_2^-, \tilde{\psi}_2). \tag{7.31}$$

Consequently, with Proposition 7.9 at hand, we get

$$(\Phi^{-1})^* \left(\frac{D}{dt} \Big|_{t=1} \dot{\sigma}_2^0(t; \tilde{\psi}_0, \chi_1^-, \tilde{\psi}_1) \right) = \frac{D}{dt} \Big|_{t=1} \dot{\gamma}_2^0(t; \tilde{p}_0, q_1^-, \tilde{p}_1), \tag{7.32}$$

and

$$(\Phi^{-1})^* \left(\frac{D}{dt} \Big|_{t=0} \dot{\sigma}_3^1(t; \tilde{\psi}_1, \chi_1^+, \chi_2^-, \tilde{\psi}_2) \right) = \frac{D}{dt} \Big|_{t=0} \dot{\gamma}_3^1(t; \tilde{p}_1, q_1^+, q_2^-, \tilde{p}_2), \tag{7.33}$$

which completes the proof. □

Algorithm 7.1 summarizes all the necessary steps for constructing a C^2 solution on $\mathcal{P}$.

Algorithm 7.1 C^2 solution on $\mathcal{P}$.

Require: $N \geq 3$, $\tilde{P} = [\tilde{p}_0, \ldots, \tilde{p}_N]^T$ a matrix of size $N(N+1) \times N$ containing the $(N+1)$ interpolation points on $\mathcal{P}$.
1:
Ensure: Q.
2:
3: Calculate $\hat{\chi} = [\hat{\chi}_1^-, \ldots, \hat{\chi}_{N-1}^-]^T$ control points of C^1 Bézier curve on $\mathcal{M}$ using Algorithm 6.
4: Set $\chi_1^- = \hat{\chi}_1^-$.
5: Calculate control point χ_1^+:
6: $\quad \chi_1^+ = \mathrm{Exp}_{\tilde{\psi}_1}(-\frac{2}{3}\mathrm{Exp}_{\tilde{\psi}_1}^{-1}(\chi_1^-))$
7: Calculate control point χ_2^-:
8: $\quad \chi_2^- = \mathrm{Exp}_{\chi_1^+}\left(\frac{1}{3}\left((d\varphi_{\tilde{\psi}_1})_{\chi_1^-}\left(\dot{\alpha}(1, \tilde{\psi}_0, \chi_1^-)\right) - 4\dot{\alpha}(0, \chi_1^-, \tilde{\psi}_1)\right)\right)$
9: **for** $i = 2 : N - 2$ **do**
10: $\quad \chi_i^+ = \mathrm{Exp}_{\tilde{\psi}_i}(-\mathrm{Exp}_{\tilde{\psi}_i}^{-1}(\psi_i^-))$
11: $\quad \chi_{i+1}^- = \mathrm{Exp}_{\psi_i^+}\left(\left((d\varphi_{\tilde{\psi}_i})_{\chi_i^-}\left(\dot{\alpha}(1, \chi_{i-1}^+, \chi_i^-)\right) - 2\dot{\alpha}(0, \chi_i^-, \tilde{\psi}_i)\right)\right)$
12: **end for**
13: Calculate control point χ_{N-1}^+:
14: $\chi_{N-1}^+ = \mathrm{Exp}_{\tilde{\psi}_{N-1}}(-\frac{2}{3}\mathrm{Exp}_{\tilde{\psi}_{N-1}}^{-1}(\chi_{N-1}^-))$
15: **for** $i = 0 : N$ **do**
16: $\quad$ Calculate control point q_i control points of C^2 Bézier curve on $\mathcal{P}$:
17: $\quad q_i^- = \Phi^{-1}(\chi_i^-)$ and $q_i^+ = \Phi^{-1}(\chi_i^+)$
18: **end for**
19: **return** Q.

7.4 Experiments

Here, we use details from the previous sections to interpolate different probability density functions p_i. For each example, we present the optimal solution as C^1 and C^2 interpolating splines using Algorithm 7.1.

We consider general and nontrivial numerical examples where probability density functions are not necessarily parametric: The only conditions are non-negativity

7.4 Experiments

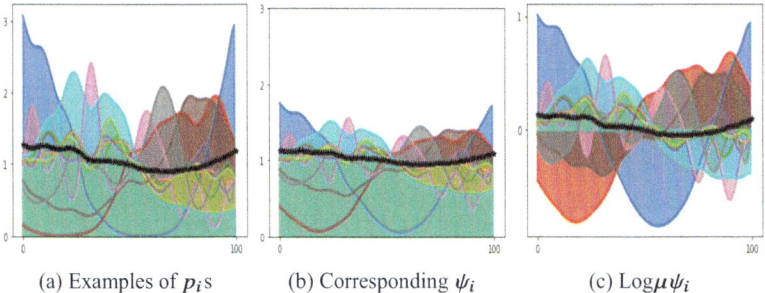

(a) Examples of p_is (b) Corresponding ψ_i (c) Log$_\mu \psi_i$

Fig. 7.3 Some examples of probability density functions p_is used for illustrations **a**, their corresponding square root ψ_is and the mean μ **b** and $\text{Log}_\mu \psi_i$. For each case, the corresponding mean is plotted as a black dotted line

with an integral equal to one. So we can not restrict applications to a specific family of PDFs. We display 50 examples of PDFs in Fig. 7.3a with different colors. In all experiments, we sample the original functions using 100 equally spaced points in $I = [0, 1]$. We illustrate some examples and their mean in Fig. 7.3a. From Fig. 7.3a, examples are different which capture large variability for illustration. This is confirmed with the corresponding mean where the displayed line is close to 1 considered as the theoretical mean when the number (m) of probability density functions $p_i, i = 1, \ldots, m$ tends to infinity. Moreover, Fig. 7.3b illustrate the corresponding square density functions ψ_i and their mean μ whereas Fig. 7.3c shows the corresponding tangent vectors $\text{Log}_\mu \psi_i$. Theoretically, the mean of $\text{Log}_\mu \psi_i$ should be zero when the number of examples tends to infinity. We can observe that the numerical solution provides a good approximation.

A first example of C^1 and C^2 interpolating splines between 5 probability density functions is given in Fig. 7.4. On the top row, we display data points denoted by p_1, p_2, p_3, p_4 and p_5 respectively. On the middle row (g), we show the solution as a C^2 interpolating spline, where elements on the diagonal are control points. The PDFs in between are intermediate points taken uniformly from the resulting solution. Note that we can generate as many intermediate points as we want, but we restrict ourselves to 5 in order to provide very qualitative illustrations. Indeed, using more points will decrease size of sub-figures and lead to hard comparisons between C^1 and C^2 solutions. We use the same idea to illustrate the C^1 interpolating spline.

To demonstrate applications of the proposed methods, we consider different examples in Figs. 7.4, 7.5, 7.6 and 7.7. We first consider examples that are quite close or standard PDFs followed by more complicated ones. In each example, we show control point on the top followed by C^2 solution in the middle the C^1 solution at the bottom. Colors are only used to distinguish different and uniformly temporal instants in I: From light (p_0 at $t = 0$) to dark ($p_n = p_5$ at $t = 1$). In all experiments, one can observe, at least visually, that the C^2 solution is more smooth around the control points. This is confirmed by computing the vector field $\dot{\sigma}$ from each solution and for each example. All results are summarized in Fig. 7.8.

130   7 Spline Interpolation on the Manifold of Probability Density Functions

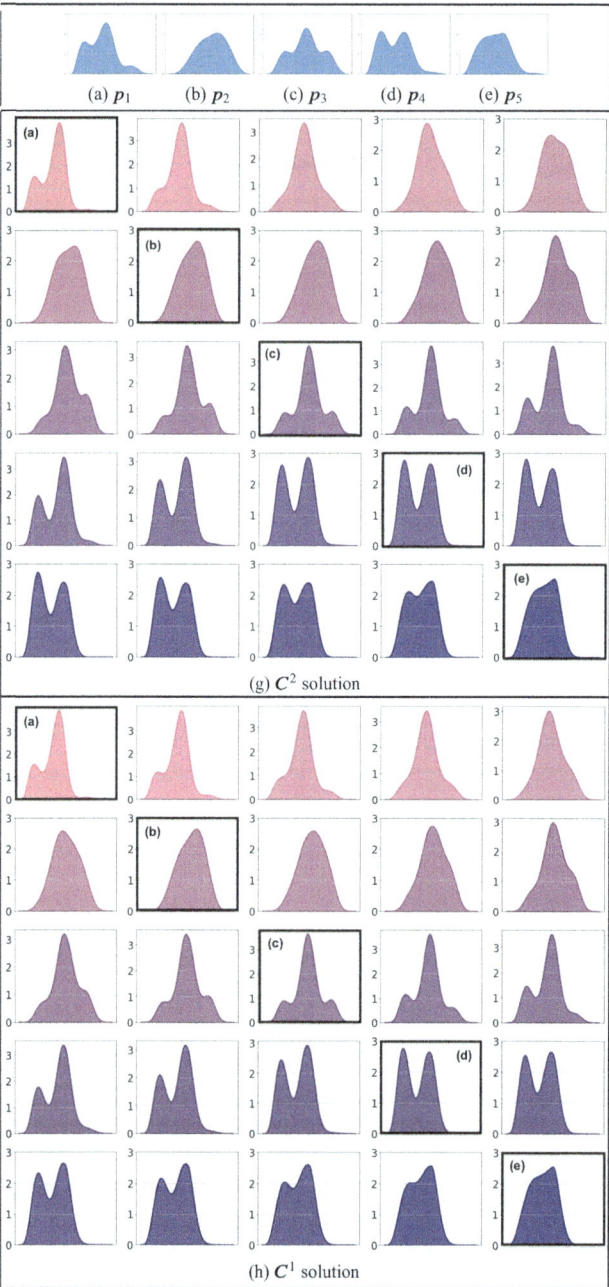

Fig. 7.4 Interpolating p_1, p_2, p_3, p_4 and p_5 with C^1 and C^2 solutions

7.4 Experiments

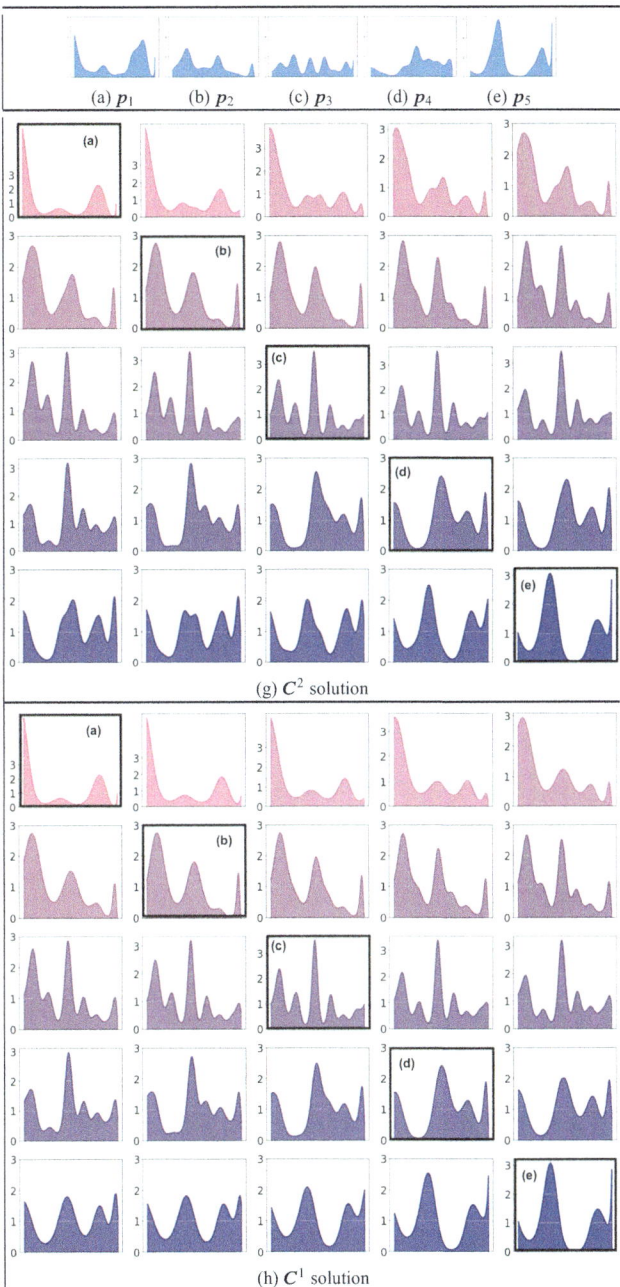

Fig. 7.5 Interpolating p_1, p_2, p_3, p_4 and p_5 with C^1 and C^2 solutions

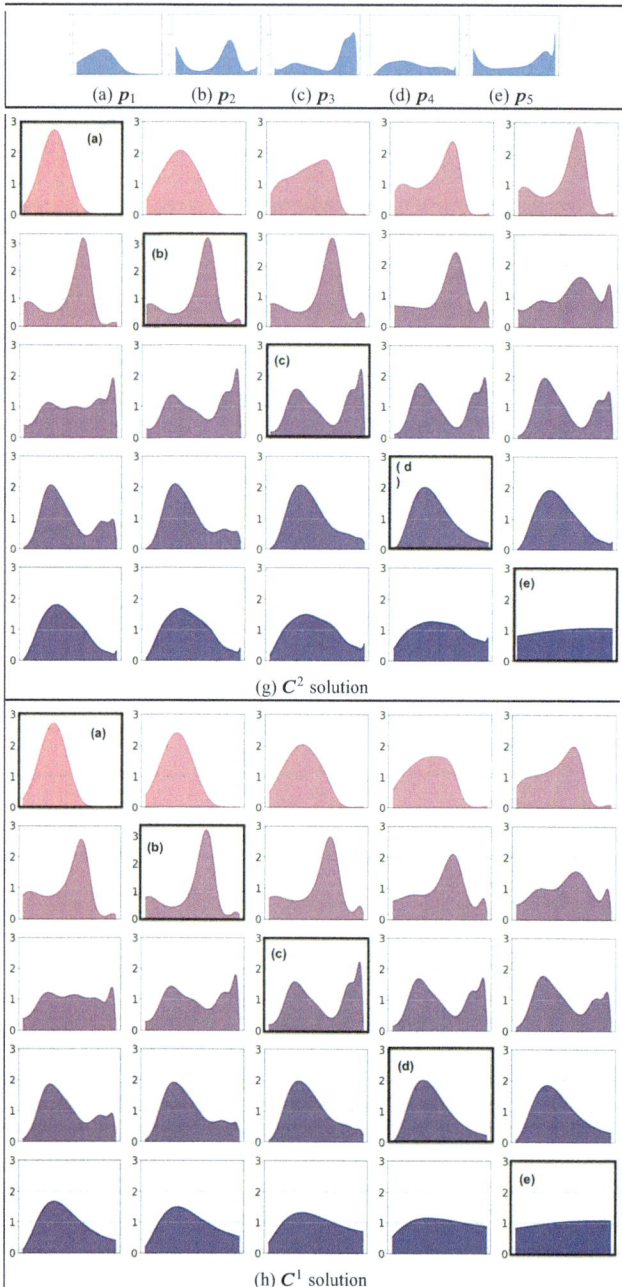

Fig. 7.6 Interpolating p_1, p_2, p_3, p_4 and p_5 with C^1 and C^2 solutions

7.4 Experiments

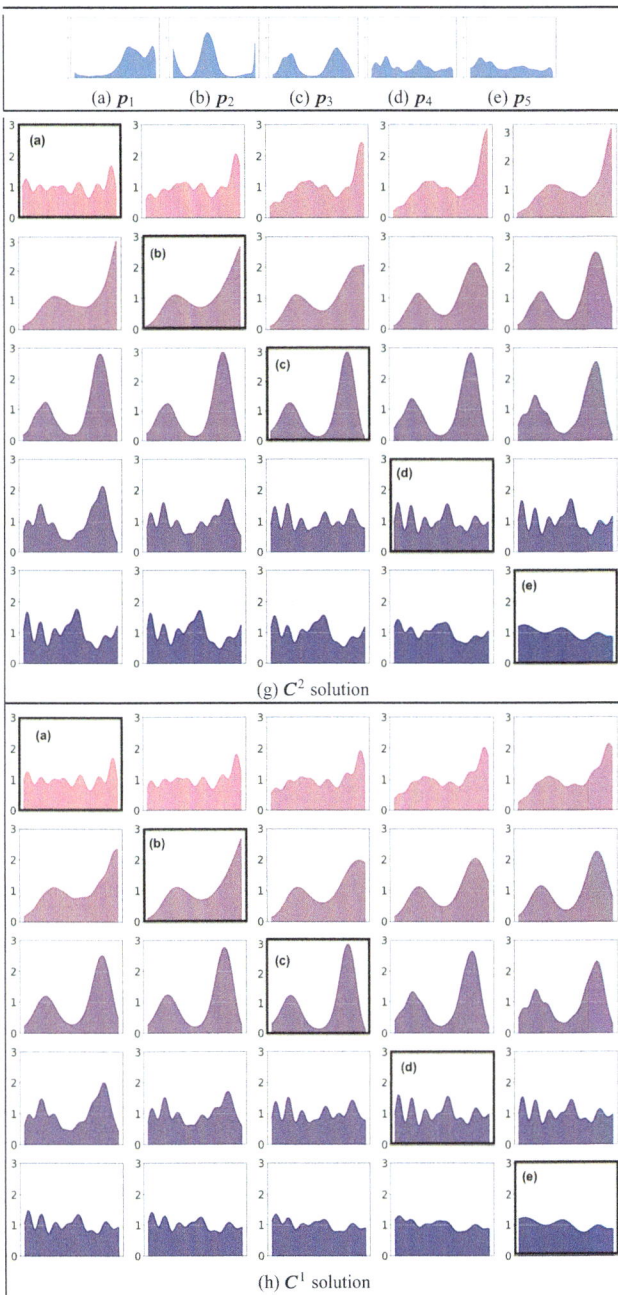

Fig. 7.7 Interpolating p_1, p_2, p_3, p_4 and p_5 with C^1 and C^2 solutions

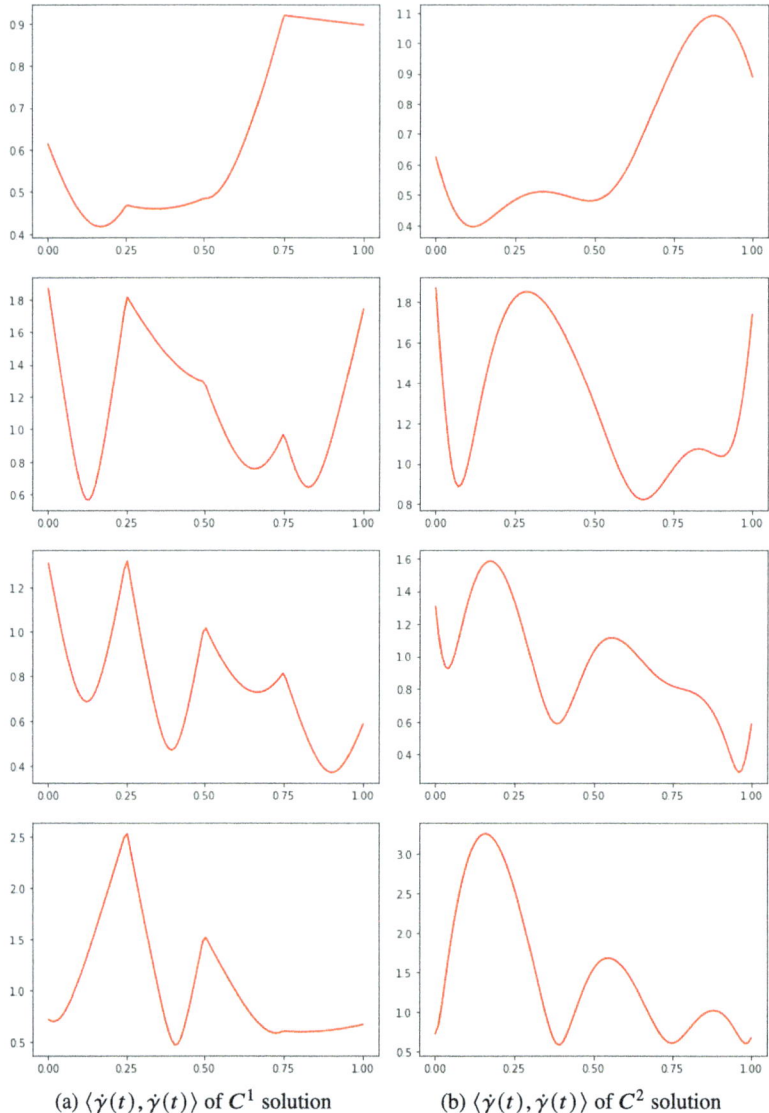

Fig. 7.8 The squared norm of the vector field $\dot{\gamma}$. The first derivative of C^1 solution (**a**) and the first derivative of C^2 solution (**b**). From top to bottom are examples from Figs. 7.4, 7.5, 7.6 and 7.7), respectively

References

1. Buckland, S.T.: Fitting Density Functions with Polynomials, Journal of the Royal Statistical Society, 41 (1), 63-76, 1992.
2. Clavijo-Blanco, J.A. and González Cagigal, M.A. and Rosendo-Macías, J.A.: A fitting procedure for probability density functions of service restoration times. Application to underground cables in medium-voltage networks, Electric Power Systems Research, 217, 2023.
3. Ang, V.: Financial interaction analysis using best-fitted probability distribution, Indicators to support monetary and financial stability analysis: data sources and statistical methodologies, 39, 2015.
4. Le Borne, S. and Wende, M.: Iterative Solution of Saddle-Point Systems from Radial Basis Function (RBF) Interpolation, 41(3), A1706–A1732, 2019.
5. Celledoni, E. and Eidnes, S. and Owren, B. and Ringholm, T.:Dissipative Numerical Schemes on Riemannian Manifolds with Applications to Gradient Flows, SIAM Journal on Scientific Computing, 40(6), A3789–A3806, 2018.
6. Bardsley, J.M. and Cui, T. and Marzouk, Y. M. and Wang, Z.: Scalable Optimization-Based Sampling on Function Space, SIAM Journal on Scientific Computing, 42(2), A1317–A1347, 2020.
7. Ay, N. and Jost, J. and Le, H.V and Schwachhofer, L.: Information geometry, Springer, Cham, Switzerland, 2017.
8. Friedrich, T.: Die Fisher-information und symplektische strukturen, Mathematische Nachrichten, 153, 273–296, 1991.
9. Srivastava, A. and Klassen, E.: Functional and Shape Data Analysis, Springer, New York, NY, 2016.
10. Rao, C.R: Information and accuracy attainable in the estimation of statistical parameters, Bull. Calcutta Math. Soc, 37, 81–89, 1945.
11. Cencov, N.N: Statistical decision rules and optimal inference, Translations of Mathematical Monographs, 53, 81–89, 1982.

Chapter 8
Spline Interpolation on Shape Space

For many applications in several branches of science, involving medical imaging, computer vision, Human Biometrics and Nanomanufacturing, it is desirable to be able to characterize objects for detection, recognition and prediction of their behavior at unobserved times or in the future. For instance, when quantifying growths and decays of tumors after medical diagnosis using medical images, an algorithm for analyzing shapes can help to identify changes in organs and consequently deduce the progression of diseases, see [1] for more details.

Statistical analysis and modeling of shapes of objects take their origin in works established by Kendall [2] where the shape of an object in Euclidean space is defined as all the geometrical information that remains when location, scale and rotational effects are filtered out from an object. Many different methods for fitting geodesics in Kendall shape space have been proposed. Le and Kume [3] fitted geodesics in planer Kendall shape space. Kenobi and al. [4] investigated a method for fitting minimal geodesics in Kendall shape space based on minimizing sums of squares of Procrustes distance. The authors of [5] propose a method for fitting smoothing splines in general Kendall shape space using the result obtained in Jupp and Kent [6] for unrolling and unwrapping procedures in order to fit spherical splines to spherical data.

While Kendall's definition of shape space took major strides in shape analysis, it admits some limitations due to the use of landmarks to define shape space. Therefore, much work has been done in order to find a convenient representation of shapes that enables simple physical interpretations of deformations of shapes and efficient method for fitting curves and geodesics. Srivastava et al. [7, 8] propose a new geometric representations of curves based on computational differential geometry. First, they were restricted to arc-length parametrization of curves [9]. Then, they propose a square-root velocity function (SRVF) for analyzing shapes of curves in Euclidean spaces under an elastic metric and compute geodesics between closed curves using path-straightening approach [7]. Other authors have also presented other alternative methods in order to find optimal curves in shape spaces [10–12].

This chapter addresses the problem of fitting a C^1 Bézier path to a given finite set of ordered data points on the shape space of curves. We establish the existence and uniqueness properties of the path and introduce a numerical method for constructing an optimal solution. The proposed method incorporates local elastic geodesic deformation vector fields between successive curves to estimate global deformation. We showcase the effectiveness of the method in predicting missing data from ordered observations in various medical examples.

8.1 Problem Formulation

Let I be the unit interval and let $\mathbb{L}(I, \mathbb{R}^3)$ be the set of square integrable functions from I to $\mathbb{R}^3$. Let $\eta : I = [0, 1] \longrightarrow \mathbb{R}^3$ denotes a parametrized curve in $\mathbb{L}(I, \mathbb{R}^3)$ satisfying: (i) η is absolutely continuous (ii) $\dot{\eta} \in \mathbb{L}(I, \mathbb{R}^3)$. Note that absolute continuity is equivalent to requiring that $\dot{\eta}(t)$ exists for almost $t \in I$, that $\dot{\eta}(t)$ is summable and that $\eta(t) = \int_0^t \dot{\eta}(s) ds$. Now, we shall take an ordered set of open curves $p_0, \ldots, p_N$ satisfying conditions described above and consider the shape representation of each $p_i, i = 0, \ldots, N$ by their q-functions $q_0, \ldots, q_N$. Let $\mathcal{M}$ denote the space of these shapes. Indeed, $\mathcal{M}$ is an infinite-dimensional manifold with a Riemannian structure on it as will be detailed in Sect. 9.2. Given an increasing sequence of time instants $t_0, \ldots, t_N$, the goal is to find a smooth path $\gamma : [t_0, t_N] \to \mathcal{M}$ based on generalized Bézier curve which optimally interpolate shapes ($\gamma(t_i) = q_i$). Since $\mathcal{M}$ is not a linear manifold, it is not immediately obvious that there is a sensible way to define this path.

Methods for reconstructing objects from a finite collection of section contours have been studied for decades and have been successfully applied in different areas of research including computer graphics, computer-aided design and geometric modeling. The major factor affecting their accuracy is related to the parametrization of the section curves. For many years, several authors have adopted a re-parametrization of cross-sectional curves based on nonrational B-splines representation [13–15]. Although these methods seem computationally efficient, the independent approximation of curves would produce inconsistent parameterizations and the eruption of control points in the longitudinal surface direction [16]. Alternative solutions include B-spline skinning [17], contour-based approaches [18, 19], etc. However, they perform weakly when adjacent curves represent high geometric variability [20, 21].

The proposed method is based on studying the geometry and its changes of curve shapes and thus differs from the skinning and the contour-based methods. In this respect, our modeling and design paradigm is entirely geometric. We present a novel framework to treat path of shapes in the setting of Riemannian geometry and Bézier curve approach. Indeed, working in such Riemannian manifold allow nice properties to solve the issues above. In the proposed approach, we show that the problem of parametrization of the section curves, unsolved with the skinning method, is solved which allow us to formulate the surface reconstruction from curves to constructing a smooth path between them on the shape space. Roughly speaking, the given curves

η_i, are represented by their corresponding shapes q_i on a nonlinear manifold $\mathcal{M}$ and a généralized Bézier path γ is used to interpolate them as control points on $\mathcal{M}$. In this context, we would point out that the formulation problem is well posed as an optimization cost guaranteeing an optimal solution.

In this chapter, we shall show that it makes sense to define a C^1 Bézier path on the shape space of curves $\mathcal{M}$ as the result of a least squares minimization. The proposed method of calculating the interpolating C^1 path will be shown to enjoy a number of nice properties: it is a natural analogue of the Bézier based fitting in Euclidean spaces which could be considered as a generalization and the solution exists and is unique in many common situations. We describe efficient numerical methods to this end and apply it for real applications in medical studies. As far as we know, this kind of fitting on shape space has not been considered in the past.

8.2 Geometry of Shape Space

In this section we adapt the ideas from [7, 8] to study the geometry of shapes. In fact, we are interested in the q-function representation of continuous curves in $\mathbb{R}^3$. It is shown that is an efficient representation for analyzing shapes of curves. Furthermore, it is the representation in which the elastic metric reduces to a simple L^2 metric and the space of unit length curves becomes the unit Hilbert sphere.

8.2.1 Curve Representation and Shape Space

Let $\eta : I = [0, 1] \longrightarrow \mathbb{R}^3$ denote a parametrized curve in $\mathbb{L}^2(I, \mathbb{R}^3)$ satisfying conditions described in the previous section. For the purpose of studying its shape, we will represent it using its q-function $q : I \longrightarrow \mathbb{R}^3$ defined as:

$$q(s) = \frac{\dot{\eta}(s)}{\sqrt{||\dot{\eta}(s)||}} \in \mathbb{R}^3. \tag{8.1}$$

Here $\|.\|$ denotes the Euclidean 2-norm in $\mathbb{R}^3$. This vector valued function q is the tangent vector normalized by the square-root of the instantaneous speed along the curve and is a local descriptor of the geometry of the curve. The original curve η can be reconstructed using $\eta(s) = \int_0^s ||q(t)|| \, q(t) \, dt$. The scale invariant shape representation is given by normalizing the function q by its magnitude as $\frac{q}{\sqrt{\int_0^1 ||q(s)||^2 ds}}$.

Therefore, it becomes an elements of a unit sphere in the Hilbert manifold $\mathbb{L}^2(I, \mathbb{R}^3)$ that we will denote $\mathcal{M}$. This is an infinite-dimensional Hilbert manifold and represents the shape space of all translation and scale-invariant open elastic curves. More precisely, we can define the manifold $\mathcal{M}$ in the form:

$$\mathcal{M} \equiv \left\{ q \in \mathbb{L}^2(I, \mathbb{R}^3) | \int_0^1 (q(s), q(s))_{\mathbb{R}^3} ds = 1 \right\}. \tag{8.2}$$

Here and subsequently, $(,)_{\mathbb{R}^3}$ stands for the standard Euclidean inner-product in $\mathbb{R}^3$. Fitting data on the shape space $\mathcal{M}$ is the subject of this chapter.

8.2.2 Geodesics in Shape Space, Exponential Map and Parallel Transport

An important geometrical construct for the statistical analysis of shapes is the definition of the tangent space. Since $\mathcal{M}$ is a Hilbert sphere in $\mathbb{L}^2(I, \mathbb{R}^3)$, at any curve $q \in \mathcal{M}$, we define the tangent space and we denote $T_q \mathcal{M}$. We equip the tangent space of $\mathcal{M}$ with a smoothly varying Riemannian metric that measures infinitesimal lengths on the shape space. This inner product is first defined generally on $\mathbb{L}^2$ and then induced on the tangent space of $\mathcal{M}$. The metric defined on $\mathcal{M}$ has a nice physical interpretation in being an elastic metric. More precisely, let f and g be two tangent vectors in $T_q \mathcal{M}$, the metric is defined as,

$$\langle f, g \rangle = \int_0^1 (f(s), g(s))_{\mathbb{R}^3} ds. \tag{8.3}$$

Another important step in in the proposed shape analysis is to compute geodesic paths between shapes with respect to the chosen metric. With respect to the q-function, $\mathcal{M}$ is represented as the Hilbert sphere in $\mathbb{L}^2(I, \mathbb{R}^3)$ and obviously lot is known about the geometry of a sphere, including geodesics and exponential map. Therefore, geodesics between any two points q_1 and q_2 (not antipodal to q_1) on $\mathcal{M}$ are great circles and it is expressed in terms of a tangent direction $f \in T_{q_1} \mathcal{M}$ as,

$$\chi_t(q_1; f) = \cos(t\|f\|) q_1 + \sin(t\|f\|) \frac{f}{\|f\|}. \tag{8.4}$$

This equation gives the constant-speed parameterization of the geodesic passing through q_1 with velocity vector f at $t = 0$. As a result, the exponential map $\exp : T_{q_1} \mathcal{M} \longrightarrow \mathcal{M}$ is defined as

$$\exp_{q_1}(f) = q_2 = \cos(\|f\|) q_1 + \sin(\|f\|) \frac{f}{\|f\|}. \tag{8.5}$$

The length of the geodesic determines an elastic quantitative distance between two shapes q_1 and q_2 in $\mathcal{M}$ given by

$$d_Q(q_1, q_2) = \cos^{-1}(\langle q_1, q_2 \rangle). \tag{8.6}$$

8.3 Interpolation Problem on the Shape Space

From Eq. 8.4, the velocity vector along the geodesic path χ_t is obtained as $\dot{\chi}_t$. It is also noted that $\chi_0(q_1) = q_1$, and $\chi_1(q_1) = \exp_{q_1}(f) = q_2$. Conversely, given two shapes q_1 and q_2, the inverse exponential map (also known as the logarithmic map) allows the recovery of the tangent vector f between them, and is computed as follows

$$\exp_{q_1}^{-1}(q_2) = f = \frac{\cos^{-1}\langle q_1, q_2 \rangle}{\sin(\cos^{-1}\langle q_1, q_2 \rangle)} (q_2 - \langle q_1, q_2 \rangle q_1). \tag{8.7}$$

For any two points q_1 and q_2 on $\mathcal{M}$, the map $\Gamma : T_{q_1}\mathcal{M} \longrightarrow T_{q_2}\mathcal{M}$ parallel transports a vector f from q_1 to q_2 and is given by:

$$\Gamma_{q_1 \to q_2}(f) = f - 2\frac{(q_1 + q_2)\int_0^1 \langle f, q_2 \rangle ds}{\int_0^1 \langle q_1 + q_2, q_1 + q_2 \rangle ds}. \tag{8.8}$$

To summarize, the exponential map takes points in the tangent plane to points on the sphere, preserving distance from q; it also preserves the tangential direction from q. Concretely, the exponential map only preserves angles and distances for points in the tangent plane which have distance $\lesssim \pi$ from q; however, we shall implicitly assume this condition holds whenever it is needed. Given the above tools for constructing geodesics and inverse exponential maps on the shape space, we will indicate in the next section how these equations may be used to solve the problem of fitting a path to a given set of data points $q_0, \ldots, q_n$ on the shape space. We assume that all q_j are not antipodal to any $q_i, j \neq i$ or, in general, q_j are in the cut locus of q_i.

8.3 Interpolation Problem on the Shape Space

Given an ordered sequence of 3D curves represented as a set of indexed shape functions $q_0, \ldots, q_N$ in the shape space $\mathcal{M}$, the goal is to construct a smooth path $\gamma : [t_0, t_N] \to \mathcal{M}$ that interpolates q_i at t_i for $i = 0, \ldots, N$. In this context, the elastic metric will play the role of the Euclidean metric and Bézier curves of order k are generalized to $\mathcal{M}$ by means of the Riemannian de Casteljau's algorithm.

The basic idea of the introduced terminology to find the control points that generate the path γ is to treat the fitting problem on a vector space as for the Euclidean case. Clearly, the vector space is the tangent space $T_{q_i}\mathcal{M}$ at a point $q_i \in \mathcal{M}$. In order to make this possible, one needs some basic geometric ingredients. The exponential map $\exp_{q_i}$ defined on $\mathcal{M}$ by Eq. 8.5, and its inverse $\exp_{q_i}^{-1}$ given by Eq. 8.7. Actually, exponential map will transport vector fields from the tangent space $T_{q_i}\mathcal{M}$ to the shape space and conversely its inverse sends points from the manifold to the tangent space. Thus, for each $i = 1, \ldots, n - 1$, we map the rest of data points into the tangent space $T_{q_i}\mathcal{M}$ at q_i using the logarithmic map $\log_{q_i}$. The mapped data are then given by $\tilde{P} = [f_0^i, \ldots, f_N^i]$ with $f_k^i = \exp_{q_i}^{-1}(q_k)$ for $k = 0, \ldots, N$. Solving the optimization problem (2.35) in each tangent space $T_{q_i}\mathcal{M}$ provides the control point b_i^- of the Bézier curve on $\mathcal{M}$. We can now formulate the main results of this chapter.

Theorem 8.1 *Let $q_0, \ldots, q_N$ be a set of indexed shape functions in the shape space $\mathcal{M}$ with q_j being in the cut locus of q_i. For each $i = 1, \ldots, N-1$, $\tilde{P} = [f_0^i, \ldots, f_N^i]$ are the corresponding mapped data in the tangent space $T_{q_i}\mathcal{M}$ at q_i defined by $f_k^i = exp_{q_i}^{-1}(q_k)$ for $k = 0, \ldots, N$. Set $t_0 < \cdots < t_N$ a sequence of time instants. Then, there exist a unique solution X_i that minimizes Eq. 2.35 in each tangent space $T_{q_i}\mathcal{M}$ and control points that generate the Bézier curve γ interpolating the points q_i at t_i are given by: $b_i^- = exp_{q_i}(\tilde{x}_i)$ where $\tilde{x}_i = \sum_{k=0}^{N} D_{ik} f_k^i$, represent the row i of X_i in $T_{q_i}\mathcal{M}$. Moreover, the Bézier path γ is C^1 on $\mathcal{M}$.*

Proof The proof closely mirrors the proof of Theorem 3.2. □

8.4 Experiments

In this section, we demonstrate the effectiveness of the suggested approach in Sect. 4.4 to fit a smooth path to a given set of time indexed curves. Original curves were manually extracted by experts using MRI techniques during investigations in the preoperative staging and diagnosis in different specialties. Some examples of extracted curves are shown in Fig. 8.1a. In all cases, apparent boundaries are used as input

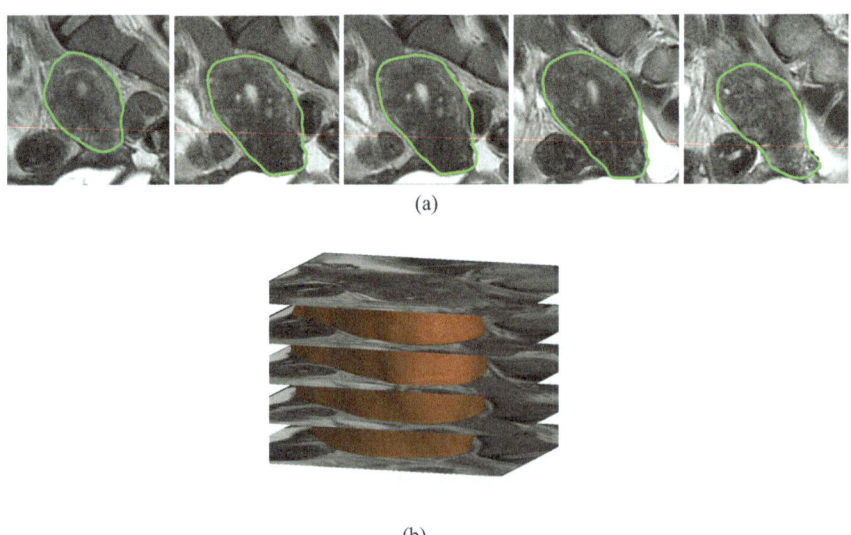

Fig. 8.1 Examples of original curves (p_i, $i = 0, \ldots, 4$) extracted from MRI images in (**a**) and MRI slices on the reconstructed path γ (**b**)

8.4 Experiments

observations to predict the missing data. Since the boundary of organs are likely tubular, it can be reconstructed as a continuous path or an evolution of the starting curve at $t = 0$. We summarize the outlined approach in Fig. 8.1 where the panel (a) displays five manually extracted curves (p_i, $i = 0, \ldots, 4$) from different MRI slices and panel (b) shows an example of the reconstructed path γ as well as the originally segmented MRI slices.

In all experiments, we initially sampled each curve with 100 points and used 50 curve points to represent the path. The quality of the fitting is measured by four criteria: the gradient field, the Laplacian field, the velocity and the acceleration. Note that we compute the geodesic between two successive curves by first finding the optimal correspondence (re-parametrization) between them, and then the geodesic deformation vector field. This allows us to efficiently capture elastic deformations along the whole path.

8.4.1 Example 1

In this section, we present the results when applying the approach presented in this chapter to fit a smooth path between curves as interpolation points at given z-levels. An illustration is given by the example displayed in Fig. 8.1a. As a first application, we performed the path fitting method as a tubular surface that we note S_{MRI}. To do so, we proceed in four steps. First, a radiologist was asked to select different slices (from 4 to 7) and segment curves as boundaries of an interest zone on each slice. Second, we represented each curve as a point q on the shape manifold $\mathcal{M}$. Note that we aligned and fixed the starting point of each curve. To make connection with time instant, we define $t_j = \frac{max(z_i) - z_j}{max(z_i) - min(z_i)} \in [0, 1]$. Third, we used a modified version of [7] to compute a geodesic segment between any two points q_i and q_j on shape space. Finally, we applied the method described in Sect. 4.4 to construct S_{MRI} as a C^1-fitting path between original curves. To give an idea about the quality of the constructed path γ, we start by showing two examples of S_{MRI} constructed from a set of 3D curves using piecewise geodesics.

8.4.1.1 A Piecewise Geodesic Path

Since we have a way to compute geodesics between any two shapes, one can connect any two successive curves at t_i and t_{i+1} by a geodesic segment. In this case, the whole interpolating path γ is a piecewise geodesic connecting the interpolation points. We display results of this piecewise geodesic path using two different examples with four original curves in Figs. 8.2 and 8.3. We can easily check that the resulting curve γ is continuous but not C^1 at the interpolation points. This leads to a distortion on the reconstructed tube as has been highlighted by the velocity and acceleration in Figs. 8.2c, d and 8.3c, d. Since γ could be seen as a tubular surface, it results that it

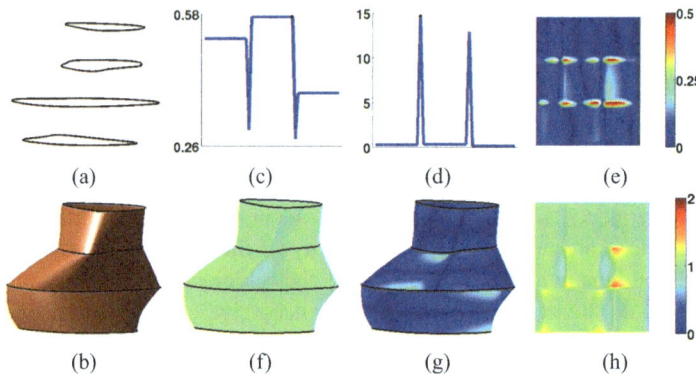

Fig. 8.2 a Original curves, **b** reconstructed surface S_{MRI}, **c** $\|\dot{\gamma}\|$, **d** $\|\ddot{\gamma}\|$, **e** $\|\nabla^2 S_{MRI}(r,\theta)\|$, **f** $\|\nabla_r S_{MRI}(r,\theta)\|$, **g** $\|\nabla_\theta S_{MRI}(r,\theta)\|$, and **h** $\|\nabla S_{MRI}(r,\theta)\|$

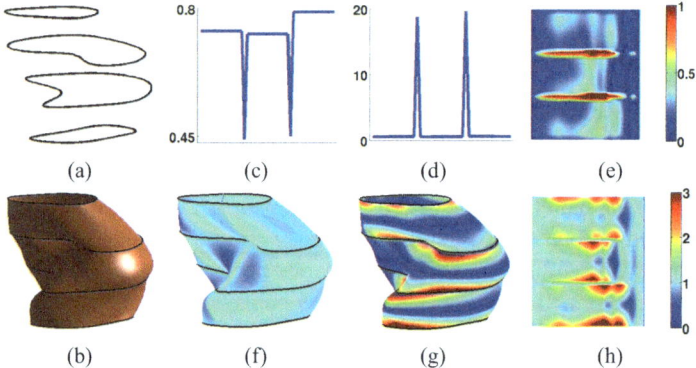

Fig. 8.3 a Original curves, **b** reconstructed surface S_{MRI}, **c** $\|\dot{\gamma}\|$, **d** $\|\ddot{\gamma}\|$, **e** $\|\nabla^2 S_{MRI}(r,\theta)\|$, **f** $\|\nabla_r S_{MRI}(r,\theta)\|$, **g** $\|\nabla_\theta S_{MRI}(r,\theta)\|$, and **h** $\|\nabla S_{MRI}(r,\theta)\|$

could be parametrized with two parameter (r, θ), where r represent time instants and θ is simply the parameter along the curve γ. For a better visualization of the path's quality, we display the gradient and the Laplacian fields in Figs. 8.2 and 8.3e, h.

8.4.1.2 Bézier Path

Following the same experimental protocol from the previous section, we show the results of the proposed method on the same two examples. Indeed, the solutions as C^1 paths is displayed in Fig. 8.4b for the first example and Fig. 8.5b for the second example. The original curves serving as the interpolation points are displayed in Figs. 8.4a and 8.5a. We can easily check that the resulting curve γ is smooth including at the interpolation points, contrary to the piecewise geodesic path. This leads to

8.4 Experiments

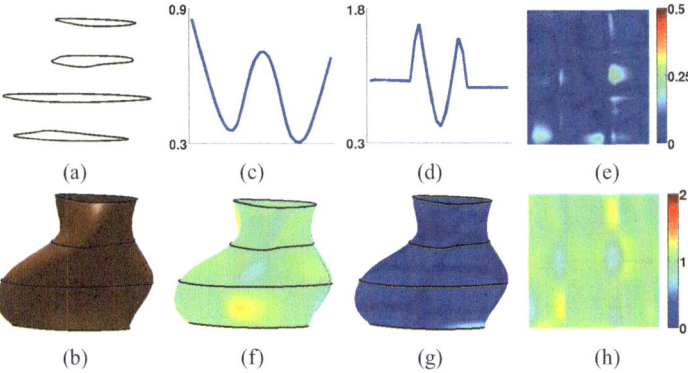

Fig. 8.4 a Original curves, b reconstructed surface S_{MRI}, c $\|\dot{\gamma}\|$, d $\|\ddot{\gamma}\|$, e $\|\nabla^2 S_{MRI}(r,\theta)\|$, f $\|\nabla_r S_{MRI}(r,\theta)\|$, g $\|\nabla_\theta S_{MRI}(r,\theta)\|$, and h $\|\nabla S_{MRI}(r,\theta)\|$

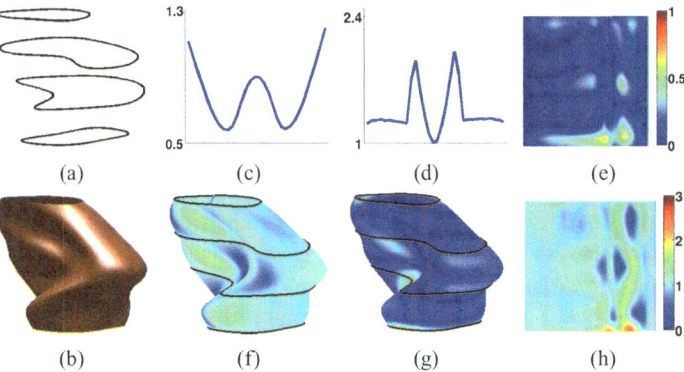

Fig. 8.5 a Original curves, b reconstructed surface S_{MRI}, c $\|\dot{\gamma}\|$, d $\|\ddot{\gamma}\|$, e $\|\nabla^2 S_{MRI}(r,\theta)\|$, f $\|\nabla_r S_{MRI}(r,\theta)\|$, g $\|\nabla_\theta S_{MRI}(r,\theta)\|$, and h $\|\nabla S_{MRI}(r,\theta)\|$

smooth reconstructed tubes as has been highlighted by the velocity and acceleration in Figs. 8.4c, d and 8.5c, d. These results are confirmed by the gradient and the Laplacian fields from Figs. 8.4e, h and 8.5e, h.

In order to validate the accuracy of the proposed framework, we present several additional real data experiments in Fig. 8.8. These examples are similar to the previous examples. We apply the proposed method to fit a smooth path to a given set of curves based on manually delineated outlines of anatomical structures. For better visualization of the smoothness of the resulting path, we show the original curves in (black) on the resulting γ.

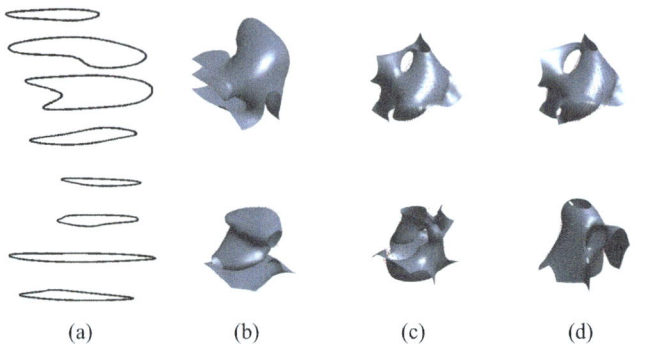

Fig. 8.6 Examples when using a polynomial fitting to reconstruct a surface. the original curves (p_i, $i = 0, \ldots, 4$) (**a**) and the reconstructed surfaces with an increasing number of intermediate curves: 0 (**b**), 5 (**c**), 10 (**d**)

8.4.1.3 Different Strategies

There is a broad range of surface reconstruction methods from a collection of contours representing cross-sections. Without being exhaustive, the goal of this textbook is to propose an alternative when some state-of-the-art methods fail to reconstruct a faithful and smooth object by ignoring the geometric changes between shapes of consecutive curves. To illustrate this idea, we show results from two different methodologies: a variational implicit function [22], a class of methods that have been detailed in [23], and an implicit polynomial fitting [24]. Both methods will be tested on the same two examples from the previous section. As confirmed by results, these two methods fail to reconstruct a good surface using only four or five indexed curves, see Fig. 8.6a for example. To improve their accuracy we increased the number of input curves using a linear interpolation in $\mathbb{R}^3$. Results for the second method are given in Fig. 8.6: (b) using 5 intermediate curves and (c) using 10 curves. We note that, even with more curves, the second method still inaccurate. On the other hand, the variational method succeeded to reconstruct one of the two surfaces in Fig. 8.7 top row. In addition, we display the mean curvature on the reconstructed surface in Fig. 8.7c to highlight the geometric quality of the reconstructed object. Nevertheless, the main problem associated with these techniques is due to an important geometric variability between neighboring curves which can not be captured if we ignore their elastic shapes. Indeed, the proposed method solves this issue by considering elastic deformations when transforming one curve to another as well as minimizing the smoothness cost.

8.4 Experiments

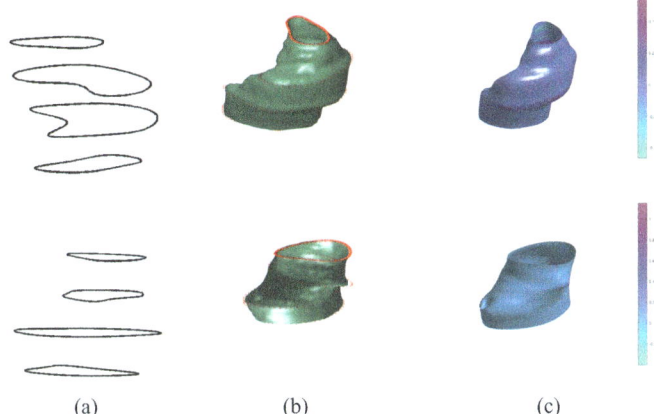

Fig. 8.7 Examples when using a variational implicit function. The original curves (p_i, $i = 0, \ldots, 4$) (**a**), the reconstructed surfaces with 10 intermediate curves (**b**), and the same surface with the mean curvature (**c**)

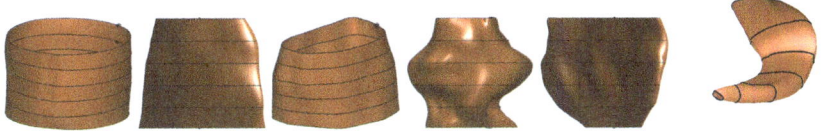

Fig. 8.8 Different examples of reconstructed γ as a C^1 Bézier path on $\mathcal{M}$ using the proposed method. The given curves are plotted in black

8.4.2 Example 2

Here, we present results using myocardial trajectories extracted as endocardial and epicardial boundaries from cine MRI images during one cardiac cycle. Even if the application is different, the goal remains the same: given different images where the zone of interest is represented by curves at ordered time instants, we want to fit a smooth path γ in order to quantify the deformation as well as the time evolution of endocardial and epicardial boundaries during the cardial cycle. By doing so, we will be able to compare two paths which is more important than just comparing static images.

References

1. Grenander, U. and Miller, M.I. and Klassen, E. and Le, H. and Srivastava, A.: Computational anatomy: an emerging discipline, Quarterly of applied Mathematics, 4, 617-694, 1998.
2. Kendall, D.G: Shape manifolds, Procrustean metrics and complex projective spaces, Bulletin of the London Mathematical Society, 16, 81–121, 1984.
3. Le, H. and Kume, A.: Detection of Shape changes in biological features, Journal of Microscopy, 200, 140–147, 2000.
4. Kenobi, K. and Dryden, I. L. and Le, H.: Shape curves and geodesic modelling, Biometrika, 97, 567–584, 2010.
5. Kim, K.R and Dryden, I.L and Le, H.: Smoothing splines on Riemannian manifolds, with applications to 3D shape space, 2018.
6. Jupp, P. E. and Kent, J. T.: Fitting smooth path to spherical data, Journal of Applied Statistics, 36, 34–46, 1987.
7. Srivastava, A. and Klassen, E. and Joshi, S. H. and Jermyn, I. H. : Shape Analysis of Elastic Curves in Euclidean Spaces, IEEE Transactions on Pattern Analysis and Machine Intelligence (PAMI), 33, 1415–1428, 2011.
8. Joshi, S.H. and Klassen, E. and Srivastava, A. and Jermyn, I.: Removing shape-preserving transformations in square-root elastic (SRE) framework for shape analysis of curves, 387–398, EMMCVPR, 2007.
9. Klassen, E. and Srivastava, A. and Mio, W. and Joshi, S.H: Analysis of planar shapes using geodesic paths on shape spaces, IEEE Transactions on Pattern Analysis and Machine Intelligence, 3 (26), 372–383, 2004.
10. Glaunes, J. and Qiu, A. and Miller, M. and Younes, L.: Large deformation diffeomorphic metric curve mapping, International Journal of Computer Vision, 80, 317–336, 2008.
11. Michor, P. W. and Mumford, D.: Riemannian Geometries on Spaces of Plane Curves, Journal of the European Mathematical Society, 8, 1–48, 2006.
12. Younes, L. :Computable elastic distance between shapes, SIAM Journal of Applied Mathematics, 58 (2), 565–586, 1998.
13. Hohmeyer, M. and Barsky, B.: Skinning rational B-spline curves to construct an interpolatory surface, Comput. Vision. Gruph. and Image Process. Graph. Mod and Image Process, 53(6), 511–521, 1991.
14. Hoschek, J.: Automatic conversion of spline curves, Comput Aided Geom, 4, 171–181, 1991.
15. Patrikalakis, N.M.: Approximate conversion of rational splines, Comput Aided Geom, 4, 155–165, 1989.
16. Piegl, L. and Tiller, W.: Surface skinning revisited, The Visual Computer, 18, 273–283, 2002.
17. Woodward, C.D.: Skinning techniques for interactive B-spline surface interpolation, Computer-Aided Design, 20(8), 441–451, 1988.
18. Keppel, E.: Approximating complex surface by triangulation of contour lines, IBM Journal of Research and Development, 19, 2–11, 1975.
19. Fuchs, H. and Kedem, Z.M. and Uselton, S.P.: Optimal surface reconstruction from planar contours, Communications of the ACM, 20(10), 1977.
20. Boissonnat, J.D.: Shape reconstruction from planar cross sections, Computer Vision, Graphics, and Image Processing, 44(1), 1–29, 1988.
21. Barequet, G. and Sharir, M.: Piecewise-linear interpolation between polygonal slices, Computer Vision and Image Understanding, 63, 251–272, 1996.
22. Salvatore C., and Ardelio G., and Giulio G., and Alfredo S.: Surface Reconstruction from Scattered Point via RBF Interpolation on GPU, CoRR, abs/1305.5179, 2013.
23. Zhao, H.K and Osher, S. and Fedkiw, R. :Fast surface reconstruction using the level set method, Proceedings IEEE Workshop on Variational and Level Set Methods in Computer Vision, 194–201, 2001.
24. Rouhani, M. and Sappa, A.D.: Relaxing the 3L algorithm for an accurate implicit polynomial fitting, 2010 IEEE Computer Society Conference on Computer Vision and Pattern Recognition, 3066–3072, 2010.

Chapter 9
Spline Interpolation on Other Riemannian Manifolds

In this chapter, our objective is to extend and validate the methodology introduced in previous chapters to encompass additional cases of symmetric Riemannian manifolds. Specifically, we focus on two such instances: the set of symmetric and positive-definite matrices (SPD), denoted as $\mathcal{P}_n^+$, and hyperbolic spaces $\mathcal{H}_n$ characterized by constant negative curvature. These nonlinear spaces find wide-ranging applications where the demand for smooth interpolating splines is pronounced. A notable application is observed in Diffusion Tensor Imaging (DTI) [1], where spatial scans are represented by 3D tensors, corresponding to 3×3 symmetric and positive-definite matrices. Interpolating DT-MRI data is pivotal for deducing fiber architecture and gaining insights into specific diseases [2]. Moreover, positive definite matrices take a key role in other contexts of tensor computing [3], materials science [4], and statistical analyses [5, 6]. This gives rise to different optimization methods that use SPD to solve PDEs and search for optimal covariance functions. Such applications and numerical approaches encourage further research for optimization methods to approximate smooth curves on Riemannian manifolds.

Conversely, hyperbolic spaces $\mathcal{H}_n$ with constant negative curvature find applications in geometry processing and machine learning. The construction of smooth interpolating curves on these spaces is essential for tasks such as shape analysis, pattern recognition, and data representation in non-Euclidean geometries. By applying the methodology to these examples, our goal is to showcase its versatility and applicability in addressing interpolation challenges across diverse domains where the underlying data conform to symmetric Riemannian manifolds.

In this chapter, we demonstrate that when $\mathcal{P}_n^+$ is equipped with the affine-invariant metric, we can define a C^2 interpolating Bézier spline as solution of the optimization problem (1.1). The same applies to $\mathcal{H}_n$ equipped with the Lorentzian inner product. In both cases, we provide explicit solutions and ensure the C^2 differentiability condition at joint points. Initially, we use the global symmetries to derive equations for control points. Subsequently, leveraging findings from previous chapters, we employ these equations to determine a C^2 interpolating Bézier spline on both Riemannian manifolds. These outcomes are adequate to yield explicit formulas for the control points of a C^2 interpolating Bézier spline on $\mathcal{H}_n$ and $\mathcal{P}_n^+$.

9.1 Spline Interpolation on Symmetric Positive Definite Matrices $\mathcal{P}_n^+$

In this section, we aim to devise a C^2 interpolating Bézier spline for the purpose of smoothing data confined to $\mathcal{P}_n^+$, the space of symmetric positive-definite matrices of order n. This space, also recognized as the set of ellipsoids in $\mathbb{R}^n$, exhibits a rich structure and several intriguing features. We commence by delving into the differential structure of $\mathcal{P}_n^+$ and providing a comprehensive derivation of the formulas governing geodesics, the Riemannian exponential map, and its inverse, all in accordance with the chosen Riemannian metric. For a more in-depth exploration of the differential geometry of $\mathcal{P}_n^+$, we refer to [7, 8].

9.1.1 Geometry of $\mathcal{P}_n^+$

Let $\mathcal{S}(n)$ denote the vector space of symmetric matrices of size n. We characterize a matrix $S \in \mathcal{S}(n)$ as positive-definite if it satisfies the quadratic form $x^T S x > 0$ for all $x \in \mathbb{R}^n \backslash \{0\}$. The set of all $n \times n$ symmetric, positive-definite matrices is denoted by $\mathcal{P}_n^+$. Using the notation $P \succ 0$, we express that P is a positive-definite matrix.

$$\mathcal{P}_n^+ = \{P \in \mathcal{M}_n(\mathbb{R}) : P^T = P \text{ and } P \succ 0\}.$$

Hence, the collection of matrices, representing the interior of an open and convex cone within the vector space of symmetric matrices $\mathcal{S}(n)$, forms a connected and differentiable manifold with a dimension of $\frac{n(n+1)}{2}$. The tangent space to $\mathcal{P}_n^+$ at any point P is defined as:

$$T_P \mathcal{P}_n^+ = \{P\} \times \mathcal{S}(n),$$

or simplicity, we identify it with $\mathcal{S}(n)$. When equipped with the standard Euclidean metric, $\mathcal{P}^n+$ forms a non-complete Riemannian manifold of non-compact type. To address this, various Riemannian metrics have been introduced to $\mathcal{P}_n^+$, with the goal of ensuring the completeness of the manifold. In this chapter, we adopt the affine-invariant Riemannian metric proposed by Siegel [9], which has been extensively studied by various authors [10, 11]. Specifically, for $P \in \mathcal{P}_n^+$ and $S_1, S_2 \in \mathcal{S}(n)$, we define the positive-definite inner product as:

$$<< S_1, S_2 >>_P = \left\langle P^{-\frac{1}{2}} S_1 P^{-\frac{1}{2}}, P^{-\frac{1}{2}} S_2 P^{-\frac{1}{2}} \right\rangle \tag{9.1}$$

$$= \text{trace}(P^{-1} S_1 P^{-1} S_2), \tag{9.2}$$

9.1 Spline Interpolation on Symmetric Positive Definite Matrices $\mathcal{P}_n^+$

that depend on the point P. A straightforward calculation reveals that this metric remains invariant under the transitive action of the general linear group $GL(n)$, defined for $L \in GL(n)$ as $P \to LPL^T$. Additionally, it is evident that the metric is also invariant under inversion ($P \to P^{-1}$). The geodesic $t \longrightarrow \alpha(t)$, satisfying $\alpha(0) = I$ and $\dot{\alpha}(t) = S \in T_I\mathcal{P}_n^+$, takes the form $\alpha(t) = \exp(tS)$. Leveraging the metric's invariance under the action of the Lie group $GL(n)$, we can derive an equation for the geodesic that meets $\alpha(0) = P \in \mathcal{P}_n^+$ and $\dot{\alpha}(t) = S \in T_P\mathcal{P}_n^+$. Explicitly, such a geodesic is given by:

$$\alpha(t) = P^{\frac{1}{2}} \exp(t P^{\frac{-1}{2}} S P^{\frac{-1}{2}}) P^{\frac{1}{2}}. \tag{9.3}$$

It follows from Hopf–Rinow theorem that $\mathcal{P}_n^+$ is a complete manifold and the Riemannian exponential map Exp_P is a global diffeomorphism and thus its inverse, denoted by Log_P, is defined in the whole space $\mathcal{P}_n^+$. Explicitly we have:

$$\mathrm{Exp}_P(S) = P^{\frac{1}{2}} \exp(P^{\frac{-1}{2}} S P^{\frac{-1}{2}}) P^{\frac{1}{2}}, \tag{9.4}$$

$$\mathrm{Log}_{P_1}(P_2) = P_1^{\frac{1}{2}} \log(P_1^{\frac{-1}{2}} P_2 P_1^{\frac{-1}{2}}) P_1^{\frac{1}{2}}, \quad P_1, P_2 \in \mathcal{P}_n^+, \tag{9.5}$$

where exp and log stands for the usual exponential and logarithm maps. Indeed, $\mathcal{P}_n^+$ serves as a homogeneous space of the Lie group $GL(n)$. Additionally, at the identity $I \in \mathcal{P}_n^+$, a symmetry function φ_I is defined by:

$$\varphi_I : \mathcal{P}_n^+ \longrightarrow \mathcal{P}_n^+, ; P \longrightarrow P^{-1}, \tag{9.6}$$

which reverses all geodesics through I. It is evident that φ_I is an isometry, as $(d\varphi_I)P$ preserves the affine-invariant Riemannian metric. Using the action of $GL(n)$, one can readily define the geodesic-reversing isometry φ at each point $P \in \mathcal{P}_n^+$. Consequently, $\mathcal{P}_n^+$ transcends being solely a Riemannian homogeneous space, it is also a Riemannian symmetric space.

The geometric toolbox Table 9.1 provides a concise overview of various geometric structures on $\mathcal{P}_n^+$.

9.1.2 C^2 Bézier Spline on $\mathcal{P}_n^+$

Given a finite set of symmetric positive-definite matrices $P_0, \ldots, P_N$ in $\mathcal{P}_n^+$ and a set of time instants $0 = t_0 < t_1 < \cdots < t_N = N$, we seek a C^2 interpolating Bézier spline $\gamma : [0, N] \to \mathcal{P}_n^+$ that minimizes the following functional

$$E(\gamma) = \int_{t_0}^{t_N} < \frac{D^2\gamma(t)}{Dt^2}, \frac{D^2\gamma(t)}{Dt^2} >_P dt \tag{9.7}$$

Table 9.1 Geometric toolbox for the special orthogonal group $\mathcal{P}_n^+$

Set	$\mathcal{P}_n^+ = \{P \in M_n(\mathbb{R}) : P^T = P \text{ and } P \succ 0\}$
Tangent spaces	$T_P \mathcal{P}_n^+ = \{P\} \times \mathcal{S}(n)$
Inner product	$<< S_1, S_2 >>_P = \text{trace}(P^{-1} S_1 P^{-1} S_2)$
Shortest geodesic	$\alpha(t) = P^{\frac{1}{2}} \exp(t P^{\frac{-1}{2}} S P^{\frac{-1}{2}}) P^{\frac{1}{2}}$
Exponential	$\text{Exp}_P(S) = P^{\frac{1}{2}} \exp(P^{\frac{-1}{2}} S P^{\frac{-1}{2}}) P^{\frac{1}{2}}$
Logarithm	$\text{Log}_{P_1}(P_2) = P_1^{\frac{1}{2}} \log(P_1^{\frac{-1}{2}} P_2 P_1^{\frac{-1}{2}}) P_1^{\frac{1}{2}}$
Geodesic symmetry	$\varphi_I : \mathcal{P}_n^+ \longrightarrow \mathcal{P}_n^+, \; P \longrightarrow P^{-1}$

Following the approach used in previous chapters, we aim to derive the equations that govern the control points of the C^2 Bézier spline γ specifically for the Riemannian manifold $\mathcal{P}_n^+$. Let's initiate the process by computing the control points $(\widehat{Z}_i^-, \widehat{Z}_i^+)$, for $i = 1, \ldots, N-1$, of the C^1 Bézier spline γ. To achieve this, our first step is to determine the matrix X^i, which contains the control point of the C^2 Bézier spline β within each tangent space $T_{P_i}\mathcal{P}_n^+$, for $i = 1, \ldots, N-1$. Once we have obtained these matrices, we can then proceed by extracting the ith row of each X^i and mapping it onto the manifold $\mathcal{P}_n^+$ using the Riemannian exponential map given by Eq. 9.4. This process yields the control points of the C^1 Bézier spline γ within $\mathcal{P}_n^+$.

Theorem 9.1 *Given $P_0, \ldots, P_N$ belonging to the complete Riemannian manifold $\mathcal{P}_n^+$ equipped with its affine-invariant Riemannian metric and $0 = t_0 < t_1 < \cdots < t_N = N$, a set of instants of times. Let $S = (S_0^i, \ldots, S_N^i)$ denote the interpolation point in $T_{P_i}\mathcal{P}_n^+$, for $i = 1, \ldots, N-1$ defined by $S_k^i = \text{Log}_{P_i}(P_k)$, for $k = 0, \ldots, N$. Then, there exists a unique matrix $X^i = [(B_1^1)^-, \ldots, (B_{N-1}^1)^-]^T$ in $\mathbb{R}^{n(N-1) \times n}$ containing the $(N-1)$ control points that generate the C^2 Bézier spline β_i, in each tangent space $T_{P_i}\mathcal{P}_n^+$ and a matrix $\tilde{S} = (\tilde{S}_0^i, \ldots, \tilde{S}_N^i)$ of size $n(N+1) \times n$ containing the new $(N+1)$ interpolation points in $T_{P_i}\mathcal{P}_n^+$ for $i = 1, \ldots, N-1$. Moreover, the rows of the matrix $\widehat{Z} = [\widehat{Z}_1^-, \ldots, \widehat{Z}_{N-1}^-]^T$ in $\mathbb{R}^{n(N-1) \times n}$, containing the $(N-1)$ control points that generate the Bézier spline γ interpolating the points P_i at t_i on $\mathcal{P}_n^+$ are given by: $\widehat{Z}_i^- = \text{Exp}_{P_i}(\tilde{x}_i)$ where $\tilde{x}_i$, represent the row i of X_i in $T_{P_i}\mathcal{P}_n^+$ and the new $(N+1)$ interpolation points in $\mathcal{P}_n^+$ are given by $\tilde{P}_k = \text{Exp}_{P_i}(\tilde{S}_k^i)$.*

Proof Proceeding similarly to the proof of Theorem 3.2 for the case $i = 1$, we can extend the same reasoning to other tangent spaces $T_{P_i}\mathcal{P}_n^+$, where $i = 2, \ldots, N-1$. To determine the control points generating the C^2 Bézier spline β on $T_{P_i}\mathcal{P}_n^+$, it is sufficient to address the optimization problem (1.1) on $T_{P_i}\mathcal{P}_n^+$. After computing the inner product of the acceleration with respect to the affine-invariant Riemannian metric defined in $\mathcal{P}_n^+$ and evaluating the integral for each term, we proceed to compute the critical points of the gradient of the energy function E. To achieve this, we introduce the cost function $f_P : T_P\mathcal{P}_n^+ \longrightarrow \mathbb{R}+$ defined by:

9.1 Spline Interpolation on Symmetric Positive Definite Matrices $\mathcal{P}_n^+$

$$f_P(X) = \langle P^{-\frac{1}{2}} X P^{-\frac{1}{2}}, P^{-\frac{1}{2}} X P^{-\frac{1}{2}} \rangle,$$

and we need to compute $Df_P(X)[H]$, the directional derivative of f_P at $X \in T_{P_1}\mathcal{P}_n^+$ in the direction $H \in T_{P_1}\mathcal{P}_n^+$. By using the two identities:

$$D(X \to AXB)(X)[H] = AHB, \ A, B \in M_n(\mathbb{R}),$$
$$D(X \to \langle f(X), g(X) \rangle)(X)[H] = \langle Df(X)[H], g(X) \rangle + \langle f(X), Dg(X)[H] \rangle.$$

We thus obtain:

$$Df_P(X)[H] = 2\langle P^{-\frac{1}{2}} X P^{-\frac{1}{2}}, P^{-\frac{1}{2}} H P^{-\frac{1}{2}} \rangle.$$

It follows that we obtain the system of equations in the Euclidean case but with respect to the affine-invariant inner product defined in $T_{P_1}\mathcal{P}_n^+$. Consequently, we can now deduce the matrix $X^1 = [(B_1^1)^-, \ldots, (B_{N-1}^1)^-]^T \in \mathbb{R}^{n(N-1) \times n}$ of size $n(N-1) \times n$ containing the control points of the C^2 Bézier spline β_1 in $T_{P_1}\mathcal{P}_n^+$. Hence:

$$(B_1^1)^- = (\widehat{B}_1^1)^-, \tag{9.8}$$

$$(B_2^1)^- = \frac{1}{3} S_0^1 - \frac{1}{2}(B_1^1)^- + \frac{8}{3} S_1^1, \tag{9.9}$$

$$(B_{i+1}^1)^- = (B_{i-1}^1)^+ + 4 S_i^1 - 4(B_i^1)^-, i = 2, \ldots, N-2, \tag{9.10}$$

and the new $(N+1)$ interpolation points $\tilde{S} = (\tilde{S}_0^1, \ldots, \tilde{S}_N^1)$ on $T_{P_1}\mathcal{P}_n^+$ are given by:

$$\tilde{S}_k^1 = S_k^1, \text{ for } k = 0, .., N-1, \tag{9.11}$$

$$\tilde{S}_N^1 = 2 S_{N-1}^1 + 2(B_{N-1}^1)^+ - 6(B_{N-1}^1)^- + 3(B_{N-2}^1)^+. \tag{9.12}$$

Taking the first row $\tilde{x}_1 = (B_1^1)^-$ of the matrix X^1 gives the first row $\widehat{Z}_1^- = \text{Exp}_{P_i}(\tilde{x}_1)$ of the matrix $\widehat{Z}$, containing the control point of the Bézier spline γ interpolating the points P_i at t_i on $\mathcal{P}_n^+$, which proves the theorem. □

Corollary 9.1 *The Bézier path* $\gamma : [0, N] \to \mathcal{P}_n^+$ *is* C^1 *on* $\mathcal{P}_n^+$.

Theorem 9.2 *Let $\tilde{P}_0, \ldots, \tilde{P}_N$ be a set of distinct points in $\mathcal{P}_n^+$ and $\alpha(t)$ the shortest geodesic arc joining control points of the spline γ on $\mathcal{P}_n^+$ given by Eq. 9.3. Let $X_1 = [(B_1^1)^-, \ldots, (B_{N-1}^1)^-]^T$ be the matrix of size $n(N-1) \times n$ containing the control points of the C^2 Bézier spline β_1 in $T_{P_1}\mathcal{P}_n^+$. Then, there exists a unique matrix $Z = [Z_1^-, \ldots, Z_{N-1}^-]^T \in \mathbb{R}^{n(N-1) \times n}$, containing the $(N-1)$ control points that generate the C^2 Bézier spline γ interpolating the points $\tilde{P}_i$ at t_i on $\mathcal{P}_n^+$, for $i = 0, \ldots, N$. The rows of Z are given by:*

(i) $Z_1^- = \text{Exp}_{\tilde{P}_1}((B_1^1)^-)$,

(ii) $Z_2^- = \text{Exp}_{Z_1^-}\left(\frac{1}{3}\left((d\varphi_{\tilde{P}_1})_{Z_1^-}\left(\dot{\alpha}(1, \tilde{P}_0, Z_1^-)\right) - 4\dot{\alpha}(0, Z_1^-, \tilde{P}_1)\right)\right),$

(iii) $Z_{i+1}^- = Exp_{Z_i^+}\left(\left((d\varphi_{\tilde{P}_i})_{Z_i^-}\left(\dot{\alpha}(1, Z_{i-1}^+, Z_i^-)\right) - 2\dot{\alpha}(0, Z_i^-, \tilde{P}_i)\right)\right)$,
$i = 2, \ldots, N - 2$.

Proof The proof is similar in spirit to the proof of Theorem 3.3. □

9.2 Spline Interpolation on Hyperbolic Space $\mathcal{H}_n$

We now explore hyperbolic spaces, characterized by a constant curvature $k < 0$, and furnish explicit formulas for computation. To avoid trivial cases, we assume $n \geq 2$. The two most commonly used hyperbolic spaces are the Poincaré ball and the hyperboloid hypersurface. Below, we present some useful definitions from hyperbolic geometry. For further elaboration, readers are directed to [12, 13]. As there is no way to differentiate between the two sheets of the hyperboloid in $\mathbb{R}^{n+1}$, we denote the positive sheet as $\mathcal{H}_n$, defined by

$$\mathcal{H}_n = \{p \in \mathbb{R}^{n+1} | <p, p> = -1, p^1 > 0\}, \tag{9.13}$$

where $<.,.>$ is the Lorentzian inner product. Thus, for any p and q we have

$$<p, q> = -p^1 q^1 + \sum_{i=2}^{n+1} p^i q^i. \tag{9.14}$$

Every Riemannian manifold isometrically diffeomorphic to $\mathcal{H}_n$ will be identified with $\mathcal{H}_n$. For $p \in \mathcal{H}_n$, the tangent space to $\mathcal{H}_n$ in p is given by

$$T_p(\mathcal{H}_n) = \{v \in \mathbb{R}^{n+1} | <p, v> = 0\}. \tag{9.15}$$

Since $<v, v> = -1$, the restriction of the Lorentzian inner product to $T_p(\mathcal{H}_n)$ is positive definite. This metric is differentiable and therefore $\mathcal{H}_n$ has a complete Riemannian structure. Starting from a point $p \in \mathcal{H}_n$ and a unit vector $v \in T_p(\mathcal{H}_n)$, we can construct a unique geodesic $\gamma : \mathbb{R} \to \mathcal{H}_n$ by the explicit formula

$$\gamma(t) = (\cosh(t))p + (\sinh(t))v \text{ for } t \in \mathbb{R}. \tag{9.16}$$

Now that we have explicit formulas for geodesics and the exponential map that guarantees the symmetry, we can generalize C^2 interpolation Bézier splines on $\mathcal{H}_n$ using the. Figure 9.1 illustrates: (a) the tangent space, tangent vectors and their corresponding geodesics, and (b&c) two different examples of C^2 interpolating splines on $\mathbb{H}^2$.

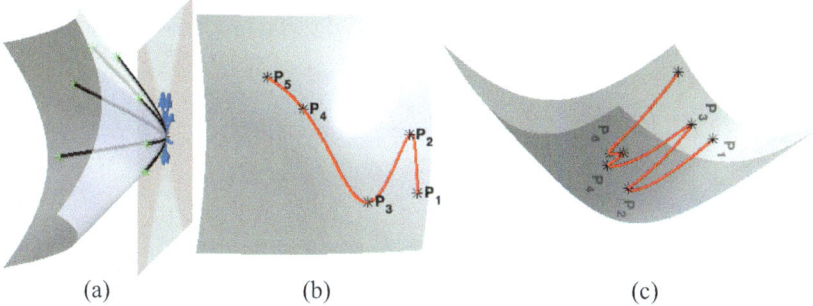

Fig. 9.1 Examples of fitting on Riemannian 2-hyperboloid. **a** Geodesics on the $\mathbb{H}^2$ and the corresponding tangent vectors. **b** and **c** are examples of interpolating C^2 splines on $\mathbb{H}^2$

References

1. Westin, C-F. and Maier, E. and Mamata, H. and Nabavi, A. and Jolesz, F.A and Kikinis, R.: Process and Visualization for Diffusion Tensor MRI, Medical Image Analysis, 6, 93–108, 2002.
2. Lazar, M. and Weinstein, D.M. and Tsuruda, J.S. and Hasan, K.M and Arfanakis, K. and Meyerand, M.E and Badie, B. and Rowley, H.A and Haughton, V.: White matter tractography using diffusion tensor deflection, Hum. Brain Mapp, 18(4), 306–321, 2003.
3. Arsigny, V. and Fillard, P. and Pennec, X. and Ayache, N.: Log-Euclidean metrics for fast and simple calculus on diffusion tensors, Journal of Magnetic Resonance In Medecine, 56(2), 411–421, 2006.
4. Sambridge, M. and Braun, J. and Mcqueen, H.: Geophysical parametrization and interpolation of irregular data using natural neighbours, Geophys. J. Int, 122(3), 837–857, 1995.
5. Moakher, M. and Zerai, M.:The Riemannian geometry of the space of positive definite matrices and its application to the regularization of positive-definite matrix-valued data, Journal of Mathematical Imaging and Vision, 40, 171–187, 2011.
6. Schwartzman, A. and Walter, F. and Jonathan, E.T:Inference for eigenvalues and eigenvectors of Gaussian symmetric matrices, Ann. Statist, 36, 2886–2919, 2008.
7. Sigurdur H.: Differential Geometry, Lie Groups, and Symmetric spaces, Academic Press, New York, 1978.
8. Terras, A.: Harmonic Analysis on Symmetric Spaces and Applications II, pringer-Verlag, New York, 1988.
9. Carl Ludwig Siegel:Symplectic Geometry, Academic Press, New York, 1964.
10. Skovgaard, L.T: A Riemannian geometry of the multivariate normal model, Scandinavian J. Statist, 11, 211–233, 1984.
11. Fletcher, P.T and Joshi, S.:Principal Geodesic Analysis on Symmetric Spaces: Statistics of Diffusion Tensors, Berlin, Heidelberg, 87–98, 2004.
12. Ratcliffe John G.: Foundations of hyperbolic manifolds, Graduate texts in mathematics, Springer, 2005.
13. John M. Lee: Riemannian Manifolds: An Introduction to Curvature, Springer Science, Graduate texts in mathematics, 1997.

Appendix A
Background Material

A.1 Basic Elements of Riemannian Geometry

This section provides a foundational overview of differential geometry, covering essential elements such as manifolds, charts, and tangent spaces. The incorporation of a Riemannian metric facilitates the measurement of lengths and angles within tangent spaces, giving rise to key notions like curve length, geodesics, and the Exponential and Log maps. These fundamental concepts establish the basis for crucial applications in fitting and interpolating manifold-valued data. Additionally, the discussion extends to more specialized topics such as Lie groups, homogeneous spaces, and symmetric spaces. A more thorough treatment of this abstract theory can be found in standard textbooks such as [1–5].

A.1.1 Charts and Manifolds

Definition A.1 Let $\mathcal{M}$ be a topological space. A **chart** on $\mathcal{M}$ is a pair (ϕ, U) where U is a subset of $\mathcal{M}$ and $\phi : U \to \phi(U)$ is a bijection from U to an open set $\phi(U) = \Omega$ in $\mathbb{R}^n$. Given $p \in U$, the elements of $\phi(p) = (x_1(p), \ldots, x_n(p))$ are called the **coordinates** of p in the chart (ϕ, U).

It's important to note that (U, ϕ^{-1}) constitutes a local parametrization. By using ϕ and ϕ^{-1} one can move between the sets U and $\phi(U)$ and perform calculations in $\mathbb{R}^n$ (Fig. A.1).

Definition A.2 Two charts (ϕ_1, U_1) and (ϕ_2, U_2) of $\mathcal{M}$ are said to be smoothly compatible if and only if $\phi_1(U_1 \cap U_2)$ and $\phi_2(U_1 \cap U_2)$ are both open in $\mathbb{R}^n$ and the **transition map**

$$\phi_1 \circ \phi_2^{-1} : \phi_2(U_1 \cap U_2) \to \phi_1(U_1 \cap U_2)$$

is a smooth diffeomorphism (Fig. A.2).

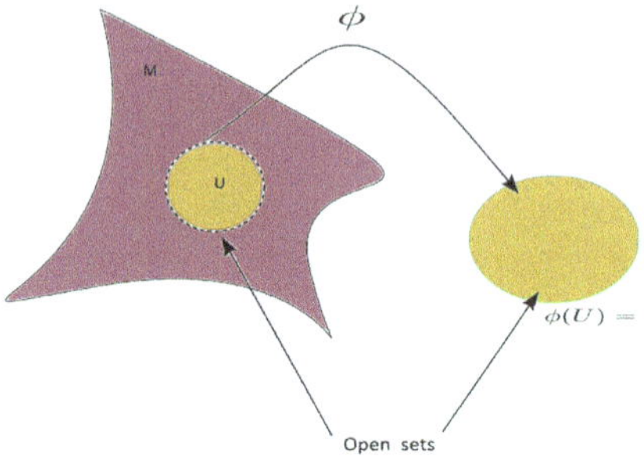

Fig. A.1 Illustration of a chart (U, ϕ) on a manifold $\mathcal{M}$

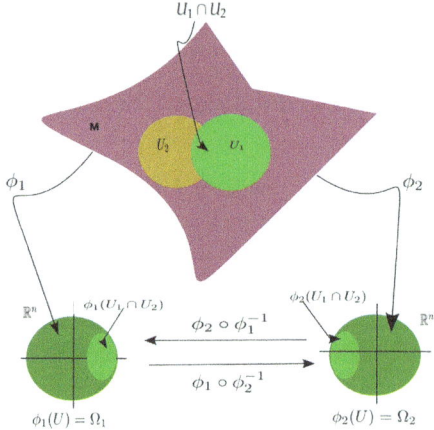

Fig. A.2 An illustration of the transition maps $\phi_1 \circ \phi_2^{-1}$ and $\phi_2 \circ \phi_1^{-1}$

Definition A.3 A **smooth atlas** on $\mathcal{M}$ is a collection $\mathcal{A} = \{(U_i, \phi_i), i \in I\}$ of charts on $\mathcal{M}$ such that any two of which are smoothly compatible and the sets U cover $\mathcal{M}$, i.e.: $\mathcal{M} = \cup_{i \in I} U_i$.

We need to ensure that we have a sufficient number of charts to effectively extend generalizing calculus from $\mathbb{R}^n$ to manifolds. To achieve this, we should have the capability to incorporate new charts whenever required, ensuring their consistency with the existing charts in an established atlas. From a technical perspective, when considering a smooth atlas $\mathcal{A}$ on $\mathcal{M}$, any additional chart (U, ϕ) is deemed compatible with $\mathcal{A}$ if and only if every mapping $\phi_i \circ \phi^{-1}$ and $\phi \circ \phi_i^{-1}$ is smooth (whenever $U \cap U_i \neq \emptyset$). Two atlases $\mathcal{A}_1$ and $\mathcal{A}_2$ are considered compatible if each chart from one atlas is compatible with the other. This condition is equivalent to stating that the union of the two atlases remains an atlas. It is readily verified that compatibility establishes an equivalence relation on $\mathcal{M}$. Indeed, when provided with an atlas $\mathcal{A}$ for $\mathcal{M}$, the collection $\tilde{\mathcal{A}}$ of all charts compatible with $\mathcal{A}$ forms a maximal atlas within the equivalence class of charts that are compatible with $\mathcal{A}$. The topology on $\mathcal{M}$ is established through the sets $(U_i)_{i \in I}$ in the maximal atlas. It is conventionally required that this topology be both Hausdorff (separable) and second countable. Nevertheless, the topology induced by the charts might not inherently possess these

Appendix A: Background Material

two properties. Additionally, when the underlying topological space of a manifold is compact, it follows that M accommodates a finite atlas.

Definition A.4 A **smooth manifold** is a pair consisting of a topological space M and a maximal smooth atlas $\mathcal{A}$ on M.

The dimension of the manifold M is n. In this manuscript, our exploration will be restricted to smooth manifolds, and henceforth, we will simply refer to them as manifolds. It is worth noting that since $\mathbb{R}^n$ is locally compact and locally connected, every manifold inherits these properties.

Example A.1 The sphere $\mathbb{S}^n = \{x = (x_0, \ldots, x_n) \in \mathbb{R}^{n+1} \mid x_0^2 + \cdots + x_n^2 = 1\}$ is an n-dimensional differentiable manifold. Specifically, $\mathbb{S}^n$ is covered with an atlas consisting of the $(2n + 2)$ charts $\phi_{i\pm} : U_{i\pm} \to \mathcal{D}^n$ where $\mathcal{D}^n$ is the open unit disk in $\mathbb{R}^n$, $U_{i\pm} = \{x \in \mathbb{S}^n \mid \pm x_i > 0\}$, and $\phi_{i\pm}$ represents the projection omitting the i th coordinate, i-e: $\phi_{i\pm}(x_0, \ldots, x_n) = (x_0, \ldots, x_{i-1}, x_{i+1}, \ldots, x_n)$.

In our applications of geometry, having only homeomorphic charts is not sufficient. We aim to be able to evaluate both first and higher derivatives of functions on our manifold. The advantageous aspect of a differentiable manifold M is that we can compute derivatives of functions on M using any coordinate chart of our choice. Moreover, when changing coordinate charts, the derivatives remain related by the chain rule.

Definition A.5 Let M and N be two smooth manifolds. A mapping $f : M \to N$ is of class C^m, if for all $p \in M$ there is a chart (U, ϕ) at p and a chart (V, φ) at $q = f(p)$, with $f(U) \subset V$ such that the map $\varphi \circ f \circ \phi^{-1} : \phi(U) \to \varphi(V)$ is of class C^m. A smooth map belongs to the class C^∞. Furthermore, if f is a bijective smooth mapping with a smooth inverse f^{-1}, it is referred to as a C^∞-diffeomorphism.

In specific cases, when considering $N = \mathbb{R}$, we obtain a C^m-function on M. From now on, $C^\infty(M)$ denotes the set of smooth functions on M.

Definition A.6 Let M be a smooth manifold, $p \in M$ and (U, ϕ) a chart at p. A **smooth parametric curve** on M is a map $\gamma : I \in \mathbb{R} \to M$ where I is some open interval of $\mathbb{R}$ such that the map $\phi \circ \gamma : I \to \phi(U)$ is a smooth function. For a closed interval $[a, b] \subseteq \mathbb{R}$, $\gamma : [a, b] \to M$ is a smooth curve from $p = \gamma(a)$ to $q = \gamma(b)$ if and only if γ can be extended to a smooth curve $\gamma : (a - \epsilon, b - \epsilon) \to M$ for some $\epsilon > 0$.

Definition A.7 Given any two points $p, q \in M$, a continuous map $\gamma : [a, b] \to M$ is a **piecewise smooth curve** from p to q if and only if

1. There is a sequence $a = t_0 < t_1 < \cdots < t_k = b$ of numbers $t_i \in \mathbb{R}$ so that each map $\gamma_i \mid [t_i, t_{i+1}]$, called a curve segment, is a smooth curve, for $i = 0, \ldots, k - 1$.
2. $\gamma(a) = p$ and $\gamma(b) = q$.

A.1.2 Tangent Spaces and Tangent Vectors

Let M be an n-dimensional manifold. There are two equivalent definitions of the tangent space T_pM at a point $p \in M$, stemming from the two equivalent definitions of the tangent vector.

Definition A.8 A **locally defined function** at p is a pair (U, f) where U is an open subset of M containing p and f is a function defined on U. Two locally defined functions, (U, f) and (V, g), at p are equivalent if and only if there is an open subset $W \subseteq U \cap V$ containing p so that $f|_W = g|_W$.

Let's denote $C_p^\infty(M)$ the set of the equivalence class of a locally defined function at p.

Definition A.9 A **tangent vector** at a point $p \in M$ is a derivation, i.e., a map $X_p : C_p^\infty(M) \to \mathbb{R}$ such that

- Linearity: $X_p(\alpha f + \beta g) = \alpha X_p(f) + \beta X_p(g)$,
- Leibniz rule: $X_p(fg) = X_p(f)g(p) + f(p)X_p(g)$,

for all $\alpha, \beta \in \mathbb{R}$ and $f, g \in C_p^\infty(M)$.

The space of all tangent vectors X_p at a point $p \in M$ is denoted as T_pM and is called the **tangent space** of the manifold M at the point p. It is easily seen that T_pM is a vector space of dimension n.

Proposition A.1 *Let M be a differentiable manifold of dimension n and $(x_1, \ldots, x_n)$ a system of local coordinates relative to a chart (U, ϕ) of M. For $p \in M$, we define $\frac{\partial}{\partial x^i}|_p \in T_pM$ par*

$$\frac{\partial}{\partial x^i}|_p : f \to \frac{\partial f}{\partial x^i}(p) = \frac{\partial (f \circ \phi^{-1})}{\partial x_i}(\phi(p)),$$

for all $f \in C_p^\infty(M)$. Here $\frac{\partial}{\partial x_i}|_p, i = 1, \ldots, n$ represents the partial derivative of a function at $p \in \mathbb{R}^n$.

There is an alternative method for defining the tangent space, particularly for a smooth manifold M embedded in $\mathbb{R}^n$. Fix a point $p \in M$ and let (U, ϕ) be a chart around p. A vector $v \in \mathbb{R}^n$ is called a tangent vector of M at p if there exists a smooth curve $\gamma : \mathbb{R} \to M$ in M such that $\gamma(0) = p$ and $\dot{\gamma}(0) = \frac{\partial}{\partial t}|_{t=0} (\phi \circ \gamma)(t) = v$. The tangent space of M at p is the linear subspace of $\mathbb{R}^n$ defined by:

$$T_pM = \{v \in \mathbb{R}^n \mid \gamma : \mathbb{R} \to M \text{ is smooth}, \gamma(0) = p, \dot{\gamma}(0) = v\}.$$

We can readily verify that, for $f \in C_p^\infty(M)$,

$$X_p(f) = \frac{\partial}{\partial t}|_{t=0} (f \circ \gamma)(t) = \frac{\partial}{\partial t}|_{t=0} f(\phi^{-1} \circ \phi \circ \gamma)(t) = \frac{\partial}{\partial t}|_{t=0} f(\phi^{-1}(\phi(p) + tv)).$$

Appendix A: Background Material

Definition A.10 Given two smooth manifolds M and N of dimensions n_1 and n_2 respectively. The **differential** of a smooth map $h : M \to N$ at p is a linear map:

$$dh_p : T_p M \to T_{h(p)} N$$

defined by

$$(dh_p(X_p))_{h(p)}(f) = X_p(f \circ h),$$

for all $X_p \in T_p M$ et $f \in C^\infty_{h(p)}(N)$.

The linear map dh_p is also denoted by Dh_p or sometimes as h_p^*.

Proposition A.2 *Let $h_1 : M \to N$ and $h_2 : N \to P$ be two smooth functions between smooth manifolds. Let $p \in M$, we have*

$$d(h_2 \circ h_1)_p = (dh_2)_{h_1(p)} \circ (dh_1)_p.$$

Example A.2 Let $h : M \to \mathbb{R}$, $p \in M$ and (U, ϕ) a chart on M at p. Since $N = \mathbb{R}$, the function $f \in C^\infty_{h(p)}(\mathbb{R})$ are just smooth functions $f : \mathbb{R} \to \mathbb{R}$ locally defined at $h(p) = t_0 \in \mathbb{R}$. Then, for any $X_p \in T_p M$, we get: $dh_p(X_p) = X_p(f \circ h)$. In particular, for $X_p = \left(\frac{\partial}{\partial x^i}\right)|_p$, we have

$$dh_p \left(\frac{\partial}{\partial x^i} \bigg|_p\right)_{t_0} (f) = \frac{\partial}{\partial x_i}(f \circ h \circ \phi^{-1})|_{\phi(p)}$$

$$= \left(\frac{\partial}{\partial x^i} \bigg|_p\right) h \frac{\partial f}{\partial t}\bigg|_{t=t_0}.$$

Hence, $dh_p(X_p) = X_p(h) \frac{\partial}{\partial t}|_{t=t_0}$.

Definition A.11 Let M be a smooth manifold. The **tangent bundle**, noted TM, is the set

$$TM = \coprod_{p \in M} T_p M = \{(p, v) \mid p \in M, v \in T_p M\},$$

where $\coprod$ stands for disjoint union. There is a natural projection $\pi : TM \to M$, where $\pi(v) = p$ if and only if $v \in T_p M$.

The smooth manifold structure of the tangent bundle is inherited from M. Furthermore, if M is a C^k manifold of dimension n, then TM is a C^{k-1}-manifold of dimension $2n$. In the case where $M = \mathbb{R}^n$, we have $TM = M \times \mathbb{R}^n = \mathbb{R}^n \times \mathbb{R}^n$.

Definition A.12 Let M be a smooth manifold. For any open subset, U of M, a **vector field** on U is a section X of TM over U, i.e., a function $X : U \to TM$ such that $\pi \circ X = Id$, (i.e., $X(p) \in T_p M$ for every $p \in U$). The vector at p is written X_p or $X(p)$ and lies in $T_p M$. The set of vector fields M is denoted as $\Gamma(TM)$.

Example A.3 Let $M = \mathbb{R}^n$ and U an open subset of M. Then, $TM = \mathbb{R}^n \times \mathbb{R}^n$ and a section of M over U is simply a function, X, such that $X(p) = (p, v)$, $v \in \mathbb{R}^n$. In other words, X is defined by a function $f : U \to \mathbb{R}^n$.

A.1.3 Riemannian Manifolds, Metrics and Isometries

Tangent spaces are vector spaces, allowing us to extend much of the comprehensive theory associated with vector spaces equipped with an Euclidean inner product to the tangent bundle of a manifold. The concept involves providing the tangent space T_pM at each point $p \in M$ on the manifold M with an inner product g_p ensuring that these inner products vary smoothly as p varies across M. This approach enables the definition of curve segment lengths on M and the establishment of distances between two points on M.

Definition A.13 Given a smooth n-dimensional manifold M, a **Riemannian metric** on M (or TM) is a collection of inner products

$$T_pM \times T_pM \to \mathbb{R} : (v, w) \to g_p(v, w)$$

such that the map

$$M \to \mathbb{R} : p \to g_p(X_p, Y_p) \tag{A.1}$$

is smooth for every pair of vector fields $X_p, Y_p \in \Gamma(TM)$. A smooth manifold M, with a Riemannian metric is called a **Riemannian manifold**.

The norm of a tangent vector $v \in T_pM$ is $\| v \|_p = \sqrt{g_p(v, v)}$. Given (U, ϕ) a chart on M at p, and $\left\{\frac{\partial}{\partial x^i}\right\}|_{i=1,\ldots,n}$ a local basis of T_pM induced by ϕ, we often use the notation $g = \sum_{ij} g_{ij} dx^i \otimes dx^j$ or simply $g = \sum_{ij} g_{ij} dx^i dx^j$, where the local expression of the metric g at p is given by the $n \times n$ matrix $g_{ij}(p)$ with

$$g_{ij}(p) = g_p\left(\left(\frac{\partial}{\partial x^i}\right)|_p, \left(\frac{\partial}{\partial x^j}\right)|_p\right).$$

The inverse is denoted by $g^{ij}(p)$. For all $p \in U$, the matrix $g_{ij}(p)$ represents a symmetric positive definite bilinear 2-form.

Example A.4 The standard Euclidean metric on $\mathbb{R}^n$, denoted as $g = dx_1^2 + \ldots + dx_n^2$, transforms $\mathbb{R}^n$ into a Riemannian manifold. Consequently, every manifold $M \subset \mathbb{R}^n$ inherits a metric by restricting the Euclidean metric to M.

Definition A.14 Given a smooth Riemannian manifold M and a smooth function $f : M \to \mathbb{R}$, the unique smooth vector field $\text{grad}(f)$ defined by

$$g_p\left((\text{grad}(f))_p, v\right) = df_p(v),$$

Appendix A: Background Material

for all $p \in \mathcal{M}$ and all $v \in T_p\mathcal{M}$ is referred to as **the gradient** of f. Thus, $\text{grad}(f) : \mathcal{M} \to T\mathcal{M}$ is a vector field on $\mathcal{M}$. Using local coordinates, we can express it as:

$$\text{grad}(f) = \sum_{ij} g^{ij} \frac{\partial f}{\partial x^i} \frac{\partial}{\partial x^j}.$$

Definition A.15 Consider two Riemannian manifolds $(\mathcal{M}_1, g_1)$ and $(\mathcal{M}_2, g_2)$. A **local isometry** is a smooth map $\varphi : \mathcal{M}_1 \to \mathcal{M}_2$ such that $d\varphi_p : T_p\mathcal{M}_1 \to T_{\phi(p)}\mathcal{M}_2$ is an isometry between the Euclidean spaces $(T_p\mathcal{M}_1, (g_1)_p)$ and $(T_{\phi(p)}\mathcal{M}_2, (g_2)_{\phi(p)})$ for all $p \in \mathcal{M}$, that is

$$(g_1)_p(u, v) = (g_2)_{\varphi(p)}(d\varphi_p(u), d\varphi_p(v)), \quad \forall u, v \in T_p\mathcal{M}_1.$$

Moreover, φ is a **global isometry**, if it is a local isometry and a diffeomorphism.

An isometry $\varphi : (\mathcal{M}, g) \to (\mathcal{M}, g)$ is referred to as an isometry of $\mathcal{M}$. The composition of isometries and the inverse of an isometry also yield isometries, forming a group known as the isometry group of $\mathcal{M}$, denoted by $\text{Isom}(\mathcal{M}, g)$. It is worth noting that $\text{Isom}(\mathcal{M}, g)$ is always a finite-dimensional Lie group that acts smoothly on $\mathcal{M}$.

Example A.5 Let $\triangle_+^n$ be the open simplex defined by

$$\triangle_+^n = \{(x_1, \ldots, x_{n+1}) \in \mathbb{R}^{n+1} \mid x_1 + \cdots + x_{n+1} = 1, \ x_i > 0\}.$$

It represents an open submanifold of the hyperplane of equation $(x_1 + \cdots + x_{n+1} = 1)$, which is itself a submanifold of $\mathbb{R}^{n+1}$. Given $x \in \triangle_+^n$, the tangent space $T_x \triangle_+^n$ is given by

$$T_x \triangle_+^n = \{u \in \mathbb{R}^{n+1} \mid u_1 + \cdots + u_{n+1} = 0\}.$$

We endow $\triangle_+^n$ with the Riemannian metric known as the Fisher information metric and defined by

$$g_x^F(u, v) = \frac{1}{4} \sum_{i=1}^{n+1} \frac{u_i v_i}{x_i}, \quad x \in \triangle_+^n.$$

The manifold $\triangle_+^n$ is diffeomorphic to the positive quadrant of the unit sphere in $\mathbb{R}^{n+1}$ given by

$$\mathbb{S}_+^n = \{(x_1, \ldots, x_{n+1}) \in \mathbb{R}^{n+1} \mid x_1^2 + \cdots + x_{n+1}^2 = 1, x_i > 0\}.$$

Indeed, the tangent space $T_x \mathbb{S}_+^n$ at $x \in \mathbb{S}_+^n$ is given by

$$T_x \mathbb{S}_+^n = \{u \in \mathbb{R}^{n+1} \mid <x, u> = 0\}$$
$$= \{u \in \mathbb{R}^{n+1} \mid x_1 u_1 + \cdots + x_{n+1} u_{n+1} = 0\},$$

where $<,>$ is the standard Euclidean inner product in $\mathbb{R}^{n+1}$. The map

$$\varphi : \mathbb{S}_+^n \to \triangle_+^n : (x_1, \ldots, x_{n+1}) \to \varphi(x_1, \ldots, x_{n+1}) = (x_1^2, \ldots, x_{n+1}^2).$$

constitutes a diffeomorphism with the inverse map given by

$$\varphi^{-1} : \triangle_+^n \to \mathbb{S}_+^n : (x_1, \ldots, x_{n+1}) \to \varphi^{-1}(x_1, \ldots, x_{n+1}) = (\sqrt{x_1}, \ldots, \sqrt{x_{n+1}}).$$

The derivative $d\varphi_x$ of φ at $x \in \mathbb{S}_+^n$ is easily verified to be

$$d\varphi_x(u_1, \ldots, u_{n+1}) = 2(x_1 u_1, \ldots, x_{n+1} u_{n+1}). \tag{A.2}$$

Consequently,

$$\begin{aligned}
g^F_{\varphi(x)}(d\varphi_x(u), d\varphi_x(u)) &= g^F_{\varphi(x)}(2(x_1 u_1, \ldots, x_{n+1} u_{n+1}), 2(x_1 v_1, \ldots, x_{n+1} v_{n+1})) \\
&= \frac{1}{4} \sum_{i=1}^{n+1} \frac{2x_i u_i 2x_i v_i}{x_i^2} \\
&= \sum_{i=1}^{n+1} u_i v_i \\
&= <u, v>
\end{aligned}$$

Hence, φ being an isometry, we can conclude that the Fisher–Rao metric is obtained as the pullback of the Euclidean metric on $\mathbb{S}_+^n$ through φ^{-1} (Fig. A.3).

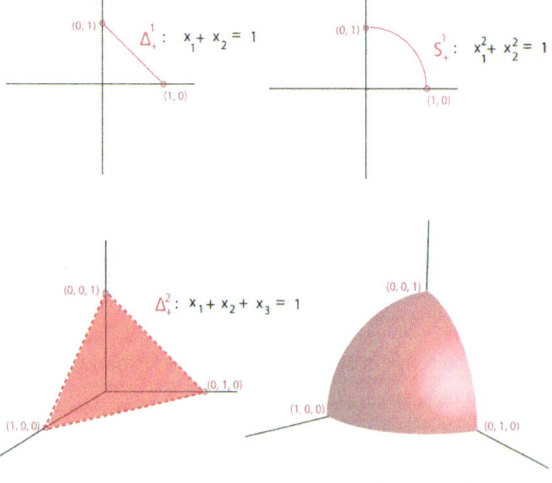

Fig. A.3 The open simplexes $\triangle_+^1$ and $\triangle_+^2$ along with the diffeomorphic $\mathbb{S}_+^1$ and $\mathbb{S}_+^2$

A.1.4 Connections and Covariant Derivative

Definition A.16 Let M be a smooth manifold, $\Gamma(TM)$ the set of smooth vector fields on M, and $C^\infty(M)$ the set of smooth function on M. A **connection** on M is an $\mathbb{R}$-bilinear map,

$$\nabla : \Gamma(TM) \times \Gamma(TM) \to \Gamma(TM) : (X, Y) \to \nabla(X, Y),$$

where we write $\nabla_X Y$ for $\nabla(X, Y)$, such that the following two conditions hold:

- $\nabla_{fX} Y = f \nabla_X Y$, C^∞-linearity in X,
- $\nabla_X (fY) = X(f) Y + f \nabla_X Y$, Leibniz rule,

for all $X, Y \in \Gamma(TM)$ and $f \in C^\infty(M)$.

A connection on M is also known as an **affine connection** on M. The notation $X(f)$ stands for a function on M such that $X(f)(p) = df(p)(X_p)$. Every smooth manifold admits many affine connections. Associated to a connection are its **torsion tensor**:

$$T(X, Y) = \nabla_X Y - \nabla_Y X - [X, Y],$$

and its **curvature** endomorphism:

$$R(X, Y) Z = \nabla_X \nabla_Y Z - \nabla_Y \nabla_X Z - \nabla_{[X,Y]} Z,$$

for all $X, Y, Z \in \Gamma(TM)$. The notation $[X, Y]$ denotes the Lie bracket of X and Y, defined as a vector field by $[X, Y] f = X(Yf) - Y(Xf)$ for all $f \in C^\infty(M)$.

Definition A.17 The vector field $\nabla_X Y$ is called the **covariant derivative** of Y with respect to X for the affine connection ∇. As $(\nabla_X Y) p \in T_p M$ depends on X only through X_p, we can interpret the notation $\nabla_v Y$, where $v \in T_p M$, as $\nabla_v Y = (\nabla_X Y)_p$ for any arbitrary $X \in \Gamma(TM)$ such that $X_p = v$.

At each point $p \in M$, the vector $(\nabla_X Y) p$ represents how the vector field Y changes at p in the direction of X_p. Consider (U, ϕ) as a chart on M at p, and let $\left\{\frac{\partial}{\partial x^i}\right\}|_{\{i = 1, \ldots, n\}}$ be a local basis of $T_p M$ induced by ϕ, we have

$$\nabla_{\frac{\partial}{\partial x^i}} \left(\frac{\partial}{\partial x^j} \right) = \sum_{k=1}^{n} \Gamma_{ij}^k \frac{\partial}{\partial x^k},$$

for some unique smooth functions, Γ_{ij}^k, defined on U, called the **Christoffel symbols**. We say that a connection ∇ is **flat** on U if and only if

$$\nabla_X \left(\frac{\partial}{\partial x^i} \right) = 0, \quad \forall X \in \Gamma(TM), 1 \leq i \leq n.$$

Definition A.18 Given a Riemannian manifold $\mathcal{M}$, there exists a unique affine connection ∇ that satisfies

- $\nabla_X Y - \nabla_Y X = [X, Y]$, **symmetry or torsion free**,
- $X g(Y, Z) = g(\nabla_X Y, Z) + g(Y, \nabla_X Z)$, **compatibility with the metric**,

for all $X, Y, Z \in \Gamma(T\mathcal{M})$.

This affine connection is called the **Levi-Civita** connection or the Riemannian connection and it can be computed using the Koszul formula:

$$2g(\nabla_X Y, Z) = X g(Y, Z) + Y g(X, Z) - Z g(X, Y) + g([x, y], Z) - g([X, Z], Y) - g([Y, Z], X).$$

Remark A.1 In a chart (U, ϕ) if we set $\partial_k g_{ij} = \frac{\partial}{\partial x_k}(g_{ij})$, then the Christoffel symbols can be expressed as

$$\Gamma^k_{ij} = \frac{1}{2} \sum_{l=1}^n g^{kl} (\partial_i g_{jl} + \partial_j g_{il} - \partial_l g_{ij}).$$

Moreover, the affine connection ∇ is torsion-free if and only if $\Gamma^k_{ij} = \Gamma^k_{ji}$.

A.1.5 Distances, Geodesic Curves and Parallel Translation

The notion of a connection gives rise to the notion of geodesic curves and parallel transport. Initially, we must establish the covariant derivative of a vector field along a curve.

Definition A.19 Let $\mathcal{M}$ be a smooth manifold and let $\gamma : I \in \mathbb{R} \to \mathcal{M}$ be a smooth curve on $\mathcal{M}$. A smooth **vector field along the curve** γ is a smooth map $X : I \to T\mathcal{M}$ such that $\pi(X(t)) = \gamma(t)$, for all $t \in I$, (i.e., $X(t) \in T_{\gamma(t)}\mathcal{M}$).

Proposition A.3 *Let $\mathcal{M}$ be a smooth manifold, let ∇ be a connection on $\mathcal{M}$ and $\gamma : I \to \mathcal{M}$ be a smooth curve on $\mathcal{M}$. There is a unique $\mathbb{R}$-linear map D/dt, defined on the vector space $\Gamma(T\mathcal{M})$ of smooth vector fields X along the curve γ which satisfies the following conditions*

1. *For any smooth function $f : I \to \mathbb{R}$*

$$\frac{D(fX)}{dt} = \frac{df}{dt} X + f \frac{DX}{dt}.$$

2. *If X is induced by a vector field $Z \in \Gamma(T\mathcal{M})$ that is $X(t_0) = Z(\gamma(t_0))$ for all $t_0 \in I$, then $\frac{DX}{dt}(t_0) = (\nabla_{\dot\gamma(t_0)} Z)_{\gamma(t_0)}$. The operator, D/dt is often called **covariant derivative along** γ and it is also denoted by $\nabla_{\dot\gamma(t)}$ or simply $\nabla_{\dot\gamma}$.*

Appendix A: Background Material

Definition A.20 Let M be a smooth manifold, ∇ a connection on M and $\gamma : I \to M$ a curve on M. A vector field X along γ is **parallel** (along γ) if and only if

$$\frac{DX}{dt}(s) = 0, \ s \in I.$$

Proposition A.4 Let $\gamma : I \to M$ be a C^1 curve on M. For every $v \in T_{\gamma(t)}M$ there is a unique parallel vector field X along γ such that $X(t) = v$.

Definition A.21 Let $\gamma : I \to M$ be a C^1 curve on M. The map

$$\Gamma_{\gamma(a) \to \gamma(t)} : T_{\gamma(a)}M \to T_{\gamma(t)}M$$

is called the **parallel transport** from $\gamma(a)$ to $\gamma(t)$ along γ.

Proposition A.5 The map $\Gamma_{\gamma(a) \to \gamma(t)}$ is a linear isomorphism, and it satisfies:

1. $\Gamma_{\gamma(a) \to \gamma(a)}$ is the identity map on $T_{\gamma(a)}M$
2. $\Gamma_{\gamma(a) \to \gamma(t)} \circ \Gamma_{\gamma(s) \to \gamma(a)} = \Gamma_{\gamma(s) \to \gamma(t)}$, $s, t \in I$.
3. The dependency of $\Gamma_{\gamma(a) \to \gamma(t)}$ on t is smooth.
4. The parallel transport associated to the Levi-Civita connection is an isometry,

$$g_{\gamma(a)}(u, v) = g_{\gamma(t)}\left(\Gamma_{\gamma(a) \to \gamma(t)}(u), \Gamma_{\gamma(a) \to \gamma(t)}(v)\right), \ u, v \in T_{\gamma(a)}M.$$

Definition A.22 Let M be a Riemannian manifold. A curve $\gamma : I \to M$ is a **geodesic** if and only if $\dot{\gamma}$ is parallel along γ. In other words,

$$\frac{D\dot{\gamma}}{dt} = \nabla_{\dot{\gamma}} \dot{\gamma} = 0.$$

If M is a Riemannian submanifold of Euclidean space $\mathbb{R}^n$, then $\dot{\gamma}$ can be replaced by γ'. Moreover, a geodesic would be a curve such that the acceleration vector $\gamma'' = \frac{D\gamma'}{dt} = 0$, i.e., γ'' is normal to $T_{\gamma(t)}M$.

Definition A.23 Let M be a Riemannian manifold. Given $p \in M$ and $v \in T_pM$, there is an interval $(-\epsilon, \epsilon)$ and a unique geodesic $\gamma_v : (-\epsilon, \epsilon) \to M$ satisfying

$$\gamma_v(0) = p, \quad \dot{\gamma}_v(0) = v.$$

The domain of γ is maximally defined; it cannot be extended further. We refer to γ_v as a **maximal geodesic**.

Theorem A.1 Consider two Riemannian manifolds (M_1, g_1) and (M_2, g_2), along with an isometry $\varphi : M_1 \to M_2$. In this case,

1. φ takes the Riemannian connection ∇^1 of M_1 to the Riemannian connection ∇^2 of M_2 in the sense that

$$d\varphi(\nabla^1_X Y) = \nabla^2_{d\varphi(X)} d\psi(Y), \quad X, Y \in \Gamma(TM).$$

2. If $\gamma : I \to M$ is a curve on M and X is a vector field along γ then the covariant derivative of X along γ is preserved, that is

$$d\varphi\left(\frac{DX}{dt}\right) = \frac{D}{dt}(d\varphi(X)).$$

3. φ takes geodesics to geodesics: if $t \to \gamma(t)$ is the geodesic on M starting from a point p with direction v, then $\varphi \circ \gamma$ is the geodesic on M with initial point $\varphi(p)$ and initial direction $d\varphi(v)$.

Definition A.24 Let M be a Riemannian manifold, and $\gamma : I \to M$ a smooth curve on M. **The length** of γ is defined by

$$L(\gamma) = \int_a^b \sqrt{g(\dot\gamma(t), \dot\gamma(t))_\gamma(t)}\, dt = \int_a^b \|\dot\gamma(t)\|_{\gamma(t)}\, dt.$$

Example A.6 1. The geodesics in $\mathbb{R}^n$ are the straight lines parametrized by constant velocity.
2. The geodesics of the 2-sphere are the great circles, parametrized by arc-length.

Definition A.25 Let M be a Riemannian manifold. A geodesic, $\gamma : I \to M$ is **minimal** if and only if its length is less than or equal to the length of any other piecewise smooth curve joining its endpoints.

Conversely, a long geodesic may not be minimal. Take, for instance, a great circle arc on the unit sphere, which is a geodesic. However, if the length of such an arc exceeds π, it is no longer minimal. Furthermore, minimal geodesics are typically non-unique. For instance, any two antipodal points on a sphere are connected by an infinite number of minimal geodesics.

Definition A.26 The Riemannian distance, also known as the **geodesic distance**, on M is defined as follows:

$$d_M : M \times M \to \mathbb{R}_+, (p, q) \to d_M(p, q) = \inf_{\gamma \in C^1(M)} L(\gamma),$$

where $C^1(M) = \{\gamma : [0, 1] \to M \mid \gamma(0) = p, \gamma(1) = q, \text{ and } \gamma \text{ is } C^1\}$.

A.1.6 Exponential and Logarithmic Maps

Definition A.27 Let M be a Riemannian manifold. For $p \in M$, consider the open subset $\mathcal{D}(p)$ of $T_p M$ defined by

$$\mathcal{D}(p) = \{v \in T_p M \mid \gamma_v(1) \text{ is defined}\},$$

where γ_v is the unique maximal geodesic with initial conditions $\gamma_v(0) = p$ and $\dot{\gamma}_v(0) = v$. The **exponential map** $\exp_p : T_p M \to M$ is given by $\exp_p(v) = \gamma_v(1)$ and

$$\exp_p(tv) = \gamma_v(t), \ tv \in \mathcal{D}(p).$$

Hence the curve $t \to \exp_p(tv)$ is the geodesic γ_v through p such that $\dot{\gamma}_v(0) = v$.

It is evident that $\mathcal{D}(p)$ is star-shaped, implying that if $w \in \mathcal{D}(p)$, then the line segment $\{tw \mid 0 \le t \le 1\}$ is entirely contained within $\mathcal{D}(p)$. In a Riemannian manifold with the Levi-Civita connection, the point $\exp_p(tv)$ is reached by traversing the geodesic γ_v for an arc length equal to $t\|v\|$ starting from p.

Definition A.28 A Riemannian manifold M is **geodesically complete** if and only if $\mathcal{D}(p) = T_p M$, for all $p \in M$, that is, if and only if the exponential $\exp_p$ is defined for all $p \in M$ and for all $v \in T_p M$.

This is equivalent to stating that the Riemannian manifold M is geodesically complete if every geodesic can be extended indefinitely.

Lemma A.1 *Let M be a Riemannian manifold. For any $p \in M$, we have: $d(\exp_p)_0 = id_{T_p M}$.*

Proposition A.6 *Let M be a Riemannian manifold. The map $\exp_p$ is a diffeomorphism from the open ball $B(0, \epsilon) \subseteq T_p M$, $\epsilon > 0$ to its image $U_p = \exp_p(B(0, \epsilon)) \subseteq M$. U_p is called a **normal neighborhood**.*

Consequently, for any $p \in M$, there is some $\beta > 0$ such that any two points $q_1, q_2 \in \exp_p(B(0, \beta))$ there is a unique geodesic from q_1 to q_2 that stays within $\exp_p(B(0, \beta))$.

Definition A.29 Let $t \to \gamma(t)$ be a geodesic curve such that $\gamma(0) = p$. We define $\gamma(t_c)$ as a cut point of this geodesic when the geodesic $t \to \gamma(t)$ ceases to be minimizing after t_c. It's worth noting that $t_c > 0$ since $\exp_p$ is a local diffeomorphism on $B(0, \epsilon) \subseteq T_p M$. The union of cut points for all geodesics starting at p is called the **cut locus** at p and is denoted by $C(p)$. Let

$$\mathcal{U}p = \{v \in T_p M \mid d_M(\exp_p(v), p) = \|v\|\}, \tag{A.3}$$

the boundary of $\mathcal{U}p$ is the cut locus at p in TpM.

Definition A.30 Let M be a Riemannian manifold. For every point, $p \in M$ the **injectivity radius** of M at p, denoted $i(p)$, is the least upper bound of the numbers, $r > 0$, such that $\exp_p$ is a diffeomorphism on the open ball $B(0, \epsilon) \subseteq T_p M$. The injectivity radius, $i(M)$ of M is the greatest lower bound of the numbers, $i(p)$ where $p \in M$.

Theorem A.2 (Hopf–Rinow) *Let M be a connected, Riemannian manifold. The following statements are equivalent:*

1. *The manifold M is geodesically complete.*
2. *For every point $p \in M$, the map $\exp_p$ is defined on the entire tangent space $T_p M$.*
3. *There is a point $p \in M$, such that $\exp_p$ is defined on the entire tangent space $T_p M$.*
4. *The metric space (M, d_M) is complete (that is, every Cauchy sequence converges).*

Definition A.31 Let M be a Riemannian manifold. We define the **Riemannian log-mapping** by the map

$$\log_p : M \to T_p M : q \mapsto \log_p(q) = v,$$

such that $\exp_p(v) = q$ and $\|v\|_p = d_M(p, q)$ (Fig. A.4).

Example A.7 The exponential map on $\mathbb{R}^n$ is given by $\exp_p(v) = p + v$, $p, v \in \mathbb{R}^n$. It is a diffeomorphism from $T_p \mathbb{R}^n = \mathbb{R}^n$ to $\mathbb{R}^n$ and hence the injectivity radius of $\mathbb{R}^n$ is infinity. Moreover, $\log_p(q) = q - p$.

Remark A.2 If a Riemannian manifold M is geodesically complete, then the geodesic distance between any two points $p, q \in M$ can be computed as $d_M = \|\log_p(q)\|_p$.

Fig. A.4 Illustration of exponential and logarithm maps

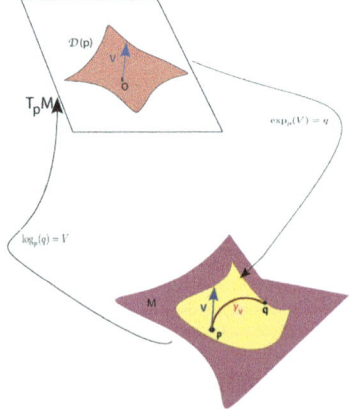

Theorem A.3 *If $\varphi : M \to N$ is a local isometry, then the following concepts are preserved:*

1. *If Γ_γ denotes the parallel transport along the curve γ and if $\Gamma_{\varphi \circ \gamma}$ denotes parallel transport along the curve $\varphi \circ \gamma$ then*

$$d\varphi_{\gamma(1)} \circ \Gamma_\gamma = \Gamma_{\varphi \circ \gamma} \circ d\varphi_{\gamma(0)}.$$

2. *If $\exp_p : T_p M \to M$ is the exponential map defined at a point $p \in M$ and $\exp_{\varphi(p)} : T_{varphi(p)} M \to N$ is the exponential map defined at a point $\varphi(p) \in N$, then we have:*

$$\varphi = \exp_{\varphi(p)} \circ d\varphi_p \circ \exp_p.$$

A.1.7 Riemannian Homogeneous Manifolds

In this section, we will introduce the concepts of homogeneous Riemannian manifold. A homogeneous Riemannian manifold is characterized by the remarkable property that its group of isometries acts transitively on the manifold, ensuring a pervasive symmetry throughout. The exploration of homogeneous Riemannian manifolds intertwines the rich theories of differential geometry and group actions. For a more comprehensive exploration of Homogeneous spaces and the theory of Lie groups, readers with an interest in the topic may consider referring [6–9].

Definition A.32 A **group** is a set G equipped with a binary operation

$$\lambda : G \times G \to G, \quad (g_1, g_2) \to g_1 g_2$$

such that

1. $g_1(g_2 g_3) = (g_1 g_2) g_3$, (associativity).
2. $ge = eg = g$, (identity).
3. There is an inversion operation map $\mu : G \to G$ that assigns to each element $g \in G$ an inverse $g^{-1} \in G$, such that $gg^{-1} = g^{-1}g = e$.

A group G is **abelian** (or commutative) if $g_1 g_2 = g_2 g_1$ for all $g_1, g_2 \in G$.

Definition A.33 A **Lie group** is a group equipped with a smooth manifold structure such that the group operations (λ and μ) are smooth maps.

Definition A.34 Given a group G, a subset H of G is a **subgroup** of G if and only if

1. $e \in H$.
2. For all $h_1, h_2 \in H$, we have $h_1 h_2 \in H$.
3. For all $h \in H$, we have $h^{-1} \in H$.

Definition A.35 H is a **Lie subgroup** of G if H is a subgroup of G and a submanifold of G.

Example A.8 1. The general linear group $GL(n, \mathbb{R})$ or simply $GL(n)$ is the set of non-singular $n \times n$ real matrices $\mathbb{R}^{n \times n}$,

$$Gl(n) = \{A \in \mathbb{R}^{n \times n} \mid \det A \neq 0\}. \tag{A.4}$$

2. A matrix Lie group is a Lie subgroup of $GL(n)$. For example, we mention:

   - **Special linear group**: $SL(n) = \{A \in GL(n) \mid \det A = 1\}$.
   - **Orthogonal group**: $O(n) = \{A \in GL(n) \mid A^T A = I_n\}$.
   - **Special orthogonal group**: $SO(n) = \{A \in O(n) \mid \det A = 1\}$.

Remark A.3 When H is a subgroup of G, and for any given $g \in G$, the set gH is called the **left coset** of H in G and is denoted by G/H. Similarly, the set Hg is called **right cosets** of H in G and are denoted by $H \backslash G$.

Definition A.36 A **compact Lie group** is a topological group whose underlying topology can be characterized as finite-dimensional locally connected compact topological groups.

Example A.9 The group of rotations $SO(n)$ is a connected, compact Lie group of dimension $d = n(n-1)/2$.

Definition A.37 Given a set $\mathcal{X}$ and a group G. A left action of G on $\mathcal{X}$ is a function $L : G \times \mathcal{X} \to G$ such that

1. For all $g_1, g_2 \in G$ and all $x \in \mathcal{X}$: $L(g_1, L(g_2, x)) = L(g_2 g_2, x)$.
2. For all $x \in \mathcal{X}$: $L(e, x) = x$.

To simplify the notation, we write $(g.x)$ instead of $L(g, x)$. Given an action $L : G \times \mathcal{X} \to G$, for any $g \in G$, we define a function $L_g : \mathcal{X} \to \mathcal{X}$ as follows: $L_g(x) = g.x$ for all $x \in \mathcal{X}$. It is evident that L_g possesses $L_{g^{-1}}$ as its inverse. Consequently, L_g is a bijection of $\mathcal{X}$ indicating that it is a permutation of $\mathcal{X}$. Particularly, if L_g is a smooth function, it is a diffeomorphism on $\mathcal{X}$.

Definition A.38 1. The **orbit** of $x \in \mathcal{X}$ under the action of G is defined as:

$$G.x = \{g.x : g \in G\} \subset \mathcal{X}$$

2. The **isotropy group** of an element $x \in \mathcal{X}$, alternatively referred to as the **stabilizer** of x, is defined as:
$$G_x = \{g \in G : g.x = x\}.$$

It is easy to verify that G_x is indeed a subgroup of G.

3. The action is said to be **free** if: $G_x = \{e\}$, $\forall x \in \mathcal{X}$.
4. The action is said to be **transitive** if: $\forall x, y \in \mathcal{X}, \exists g \in G, y = g.x$.

Remark A.4 We can also define a right action $\mathcal{X} \times G \to \mathcal{X}$ of a group G on a set $\mathcal{X}$, as a map satisfying the following conditions

1. For all $g_1, g_2 \in G$ and all $x \in \mathcal{X}$, $(x.g_1).g_2 = x.g_1g_2$.
2. For all $x \in \mathcal{X}$, $x.e = x$.

Example A.10 1. We have an action $SO(3) \times \mathbb{S}^2 \to \mathbb{S}^2$, given by: $A.p = Ap$. where Ap represents the matrix-vector product. Note that the resulting vector is still a unit vector because A is an orthogonal matrix, and it preserves the norm of the vector. This action is transitive, demonstrating that the group action can move any point to any other point on the sphere. Similarly, for any $n \geq 1$, we get a transitive action: $SO(n) \times \mathbb{S}^n \to \mathbb{S}^n$.
2. The group $Gl(n)$ acts on the set of $n \times n$ symmetric, positive, definite matrices $\mathcal{P}_+^n$ as follows: $Gl(n) \times \mathcal{P}_+^n \to \mathcal{P}_+^n$, $(A, S) \to A.S = ASA^T$. This action is transitive because every SPD matrix S can be written as $S = AA^T$, for $A \in GL(n)$.

Proposition A.7 *If the action $G \times \mathcal{X} \to \mathcal{X}$ is a transitive action of a group G on a set $\mathcal{X}$, for every fixed $x \in \mathcal{X}$ the surjection $\pi : G \to \mathcal{X}, g \to \pi(g) = g.x$ induces a bijection $\bar{\pi} : G/G_x \to \mathcal{X}$.*

In the case of a transitive action of the group G on the set $\mathcal{X}$, for any given $g \in G$, the set $\mathcal{X}$ can be expressed as a quotient (considered as a set, not as a group) of G by its isotropy subgroup G_x.

Definition A.39 A set $\mathcal{X}$ is said to be a **homogeneous space** if there is a transitive action $G \times \mathcal{X} \to \mathcal{X}$ of some group G on $\mathcal{X}$.

Theorem A.4 *Let G be a Lie group and and let H be a closed subgroup of G.*

1. *The left coset space G/H of G modulo H is a manifold of dimension equal to $(\dim G - \dim H)$, and has a unique smooth structure such that the quotient map $\pi : G \to G/H$ is a smooth submersion.*
2. *The left action of G on G/H given by: $g_1.(g_2H) = (g_1g_2)H$, $g_1, g_2 \in H$ turns G/H into a homogeneous space.*

Definition A.40 A Riemannian manifold $(\mathcal{M}, g)$ is a homogeneous if the group **Isom**$(\mathcal{M}, g)$ of its isometries is transitive on $\mathcal{M}$.

A.1.8 Riemannian Symmetric Spaces

In this section, our objective is to introduce key definitions and properties of Riemannian symmetric spaces. These spaces hold particular significance for us as they streamline the formulation of the C^2 continuity condition in Riemannian manifolds.

Elie Cartan introduced symmetric spaces and locally symmetric spaces around 1925–1930, sparking extensive research by various mathematicians. Symmetric spaces can be characterized in multiple ways, such as being simply connected Riemannian manifolds with covariantly constant curvature or Riemannian manifolds where the geodesic reflection at each point defines a global isometry. An essential implication is that symmetric spaces are homogeneous manifolds, implying that the neighborhood of any point in the space shares the same structural properties. For a more thorough examination, we refer the reader to [8, 10].

Let M be a connected Riemannian manifold. Consider $p \in M$ and U_p a normal neighborhood of p. U_p is a diffeomorphic image of a neighborhood $V = B(0, \epsilon)$ around the origin in the tangent space $T_p M$ through the exponential map $\exp_p : V \to U_p$. The **geodesic symmetry** at p is represented by the map $s_p : U_p \to U_p$, which preserves p and reflects every geodesic passing through p. In other words, it maps $\exp_p(v)$ to $\exp_p(-v)$ for every $v \in B(0, \epsilon)$.

Proposition A.8 *Let M be a connected Riemannian manifold and fix $p \in M$. The following statements are equivalent:*

1. *The geodesic symmetry s_p at p is a local isometry.*
2. *There exists a local isometry s of M defined on a neighborhood of p such that $s(p) = p$ and $ds_p = -id$.*
3. *There exists an involutive local isometry of M defined on a neighborhood of p which has p as an isolated fixed point.*

A Riemannian manifold M is called a **locally symmetric space** if the conditions outlined in Proposition A.8 are satisfied at every point in M. Moreover, M is called a **globally symmetric space**, or more simply, a symmetric space, if the geodesic symmetry s_p is globally defined on M and acts as an isometry for every $p \in M$.

Example A.11 1. The Euclidean space $\mathbb{R}^n$ is a symmetric space. The geodesic symmetry at $p \in \mathbb{R}^n$ is $s_p(q) = 2p - q$.
2. The unit sphere $\mathbb{S}^n$ in $\mathbb{R}^{n+1}$ equipped with the standard scalar product is a a symmetric space. The symmetry at any $p \in \mathbb{S}^n$ is the reflection at the line $\mathbb{R}p$ in $\mathbb{R}^{n+1}$, i.e., $s_p(q) = -q + 2 < q, p > p$.
3. Compact Lie group G with biinvariant Riemannian metric is a symmetric Riemannian manifold, where the symmetry at the unit element $e \in G$ is the inversion $s_e(g) = g^{-1}$, $g \in G$.

Another noteworthy characteristic of symmetric spaces is their possession of a parallel curvature tensor. This stems from the fact that the symmetry map s_p preserves the curvature endomorphism and its covariant derivative.

Definition A.41 A (not necessarily complete) Riemannian manifold M is a locally symmetric space if its curvature tensor is parallel, i.e., $\nabla R = 0$.

Appendix A: Background Material

Theorem A.5 $\mathcal{M}$ *is locally symmetric if and only if there exists a symmetric space $\mathcal{S}$ such that $\mathcal{M}$ is locally isometric to $\mathcal{S}$.*

In particular, if $c : \mathcal{M}_1 \to \mathcal{M}_2$ is a Riemannian covering (i.e., c is a smooth local diffeomorphism and local isometry) and $\mathcal{M}_1$ is a symmetric space, then $\mathcal{M}_2$ is locally symmetric. The manifold $\mathcal{M}_2$ does not have to be globally symmetric.

Theorem A.6 *Let $\mathcal{M}$ be a complete, simply connected locally symmetric space. Then $\mathcal{M}$ is globally symmetric.*

The converse of Theorem A.6 does not hold.

A.2 Infinite-Dimensional Riemannian Manifolds

In this section, we offer a concise introduction to infinite dimensional manifolds. A more comprehensive exploration of this topic can be found in [11, 12].

Definition A.42 A **Banach space** is a normed linear space that is a complete metric space with respect to the metric derived from its norm.

Example A.12 1. For $1 \le p < \infty$, the p-norm on $\mathbb{R}^n$ is given by

$$\|(x_1, x_2, \ldots, x_n)\|_p = (|x_1|^p + \cdots + |x_n|^p)^{1/p},$$

and for $p = \infty$, the ∞-norm by

$$\|(x_1, x_2, \ldots, x_n)\|_\infty = \max\{|x_1|, \ldots, |x_n|\}.$$

Then $\mathbb{R}^n$ equipped with the p-norm is a finite-dimensional Banach space for $1 \le p \le \infty$.

2. Let $\mathcal{F}$ be the set of all measurable functions defined on the interval $[0, 1]$. For an $f \in \mathcal{F}$ define the $\mathbb{L}^p$-norm as:

$$\|f\|_p = \left(\int_0^1 |f(x)|^p dx\right)^{1/p}.$$

With the $\mathbb{L}^p$ norm, we can define the $\mathbb{L}^p([0, 1], \mathbb{R})$ space,

$$\mathbb{L}^p([0, 1], \mathbb{R}) = \{f : [0, 1] \to \mathbb{R} \mid \|f\|_p < \infty\}.$$

$\mathbb{L}^p([0, 1], \mathbb{R})$ is a Banach space for all $p \ge 1$.

Definition A.43 A **Hilbert space** is a Banach space in which the norm is defined in terms of an inner product.

Example A.13 $\mathbb{L}^2([0, 1], \mathbb{R})$ equipped with the inner product defined by

$$< f_1, f_2 > = \int_0^1 f_1(x) f_2(x) dx, \quad f_1, f_2 \in \mathbb{L}^2([0, 1], \mathbb{R})$$

is a Hilbert space.

Definition A.44 A smooth manifold modelled on a topologival vector space E is a **Hausdorf topological space** M together with a family of charts $(U_i, \phi_i)_{i \in I}$ such that

1. $U_i \subseteq M$ are open sets, $\cup_{i \in I} U_i = M$.
2. $\phi_i : U_i \to \phi_i(U_i) \subseteq E$ are homeomorphisms onto open sets $\phi_i(U_i)$.
3. $\phi_i \circ \psi_j : \phi_i(U_i \cap U_j) \to \phi_i(U_i \cap U_j)$ are C^∞-smooth.

In this definition, the dimensionality of the space E is inconsequential, whether it is finite or infinite-dimensional. Specifically, when E is finite-dimensional, it corresponds to $E = \mathbb{R}^n$, aligning with the definition of a finite-dimensional manifold. On the other hand, if E is infinite-dimensional, the resulting manifold M is also infinite-dimensional. Various classes of infinite-dimensional E exist, encompassing Hilbert spaces, Banach spaces, Fréchet spaces, or more broadly, locally convex vector spaces. In this textbook we are interested to Hilbert Riemannian manifold.

Definition A.45 Let E be a Hilbert space. A **Riemannian Hilbert manifold** M is a smooth manifold modeled on the Hilbert space E, equipped with an inner product $< ., . >_p$ on any tangent space $T_p M$ depending smoothly on p and defining on $T_p M \cong E$ a norm equivalent to the one of E.

The principles of local Riemannian geometry for Riemannian Hilbert manifolds follow a similar path as in finite-dimensional cases. Similar to the finite-dimensional scenario, we can establish the existence and uniqueness of a symmetric connection that is compatible with the Riemannian metric. Consequently, we can define essential concepts such as covariant differentiation of a vector field along a smooth curve, parallel translation, geodesics, the exponential map, and properties like the curvature tensor R, its sectional curvature, and more, mirroring the finite-dimensional case.

Example A.14 Let E be an infinite dimensional Hilbert space equipped with an inner product $< ., . >$ and a norm $\|.\|$. The unit Hilbert sphere $\mathbb{S}^\infty$ is defined by

$$\mathbb{S}^\infty = \{f \in E \mid \|f\| = 1\}.$$

$\mathbb{S}^\infty$ is a smooth manifold. We can construct charts on $\mathbb{S}^\infty$ in the following way: For $f_0 \in \mathbb{S}^\infty$, define the subspace $E_{f_0} = \{g \in E :< g, f_0 >= 0\}$. E_{f_0} is isomorphic to E. Then define the sets $U_{f_0} = \{f \in \mathbb{S}^\infty \mid < f, f_0 > > 0\}$ and $V_{f_0} = \{g \in E_{f_0} \mid < g, g > < 1\}$. We can now define charts

$$\phi_{f_0} : U_{f_0} \to V_{f_0}, \quad f \to f - < f, f_0 > f_0$$

Appendix A: Background Material

and its inverse is given by $\phi_{f_0}^{-1}(g) = g + \sqrt{1- <g,g>} f_0$. The map

$$\phi_{h_0} \circ \phi_{f_0}^{-1}(g) = g - <g, h_0> h_0 + \sqrt{1- <g,g>}(f_0 - <f_0, h_0> h_0)$$

is a smooth isomorphism for each f_0 and h_0. In particular, the tangent space of $\mathbb{S}^\infty$ at $f \in \mathbb{S}^\infty$ is the subspace

$$T_f \mathbb{S}^\infty = \{v \in E \mid <v, f> = 0\}$$

of E with codimension one. The metric g_f for tangent vectors $u, v \in T_f \mathbb{S}^\infty$ at $f \in \mathbb{S}^\infty$ is induced from and equal to the E-inner product $<.,.>$.

In this textbook, we examine $E = \mathbb{L}^2([0,1], \mathbb{R}^n)$ endowed with the inner product $<f_1, f_2> = \int_0^1 f_1(x)^T f_2(x) dx$. The Hilbert sphere associated with this space E has been widely employed to represent density functions and shapes, extending the concept to infinite dimensions from its finite-dimensional counterpart.

References

1. Boothby, W. M.: An Introduction to differentiable manifolds and Riemannian geometry (2nd Ed), Pure and Applied Mathematics. Elsevier Science, 1986.
2. Helgason, S.: Differential Geometry, Lie Groups, and Symmetric Spaces, American Mathematical Society, 2001.
3. do Carmo, M.P.: Riemannian Geometry, Birkhäuser Boston, 2013.
4. Lee, J. M.: Introduction to smooth manifolds, Graduate Texts in Mathematics, Springer, 2002.
5. Lee, J. M.: Riemannian manifolds: An introduction to curvature, Graduate Texts in Mathematics, Springer, 1997.
6. Adams, J.F.: Lectures on Lie Groups. Midway Reprints Series. The University of Chicago Press, 1983.
7. Absil, P.A and Mahony, R. and Sepulchre,R.: Optimization Algorithms on Matrix Manifolds, Princeton University Press, 2008.
8. Arvanitogeōrgos, A.: An Introduction to Lie Groups and the Geometry of Homogeneous Spaces, AMS, 1999.
9. Jurdjevic, V.: Optimal Control and Geometry: Integrable Systems, Cambridge University Press, 2016.
10. Terras, A.: Harmonic Analysis on Symmetric Spaces and Applications II, pringer-Verlag, New York, 1988.
11. Sakai, S.: Topology of Infinite-Dimensional Manifolds, Springer, 2020.
12. Srivastava, A. and Klassen, E.: Functional and Shape Data Analysis, Springer, New York, NY, 2016.

Index

A
Acceleration, 4
Affine-invariant metric, 149
Approximating, 2

B
Bernstein polynomials, 11
Biology, 27
Biometrics, 1
Bézier curve, 1, 11
Bézier spline, 11

C
Canonical metric, 68
Cardiology, 27
Christoffel symbols, 90
Compact Lie group, 5
Composite Bézier curves, 11
Computer graphics, 50
Computer vision, 1
Constant negative curvature, 5, 149
Continuity, 9
Control points, 4, 11
Cost function, 21
Covariant acceleration, 3
Covariant derivative, 29
Cubic spline, 3, 11
    cubic polynomials, 29
Curve space, 3

D
Data modeling, 2
de Casteljau algorithm, 1

Derivative, 19
Diffeomorphism, 119
Differentiable functions, 10

E
Elastic metric, 6, 140
Energy function, 3
Euclidean metric, 68
Euclidean space, 3
Exponential map, 31
Exponential map, $\mathbb{S}^n$, 30
Exponential map, $\mathcal{P}$, 121
Exponential map, $\mathcal{P}_+(I)$, 100
Exponential map, $\mathcal{P}_n^+$, 151
Exponential map, $G_{n,p}$, 70
Exponential map, $St(n, p)$, 69
Exponential, $SO(n)$, 52

F
Fisher Rao distance, 95
Fisher-Rao metric, 5, 89, 117
Fisher-Rao metric, $\mathcal{P}$, 121
Fitting, 1
Frobenius inner product, 50

G
Geodesic, 3
Geodesic distance, 3
Geodesic distance, $\mathbb{S}^n$, 28
Geodesic distance, $\mathcal{P}$, 118, 121
Geodesic distance, $SO(n)$, 53
Geodesic regression problem, 29
Geodesic symmetry

symmetry function $\mathcal{P}_n^+$, 151
Geodesic symmetry, $\mathbb{S}^n$, 31
Geodesic symmetry, $G_{n,p}$, 70
Geodesic symmetry, $SO(n)$, 53
Geodesic, $\mathbb{S}^n$, 30
Geodesic, $\mathcal{H}_n$, 154
Geodesic, $\mathcal{P}_+$, 93
Geodesic, $\mathcal{P}_n^+$, 150
Geodesic, $G_{n,p}$, 70
Geodesic, $SO(n)$, 52
Geodesic, $St(n, p)$, 68
Geophysics, 27
Gradient, 22
Grassmann manifold $G_{n,p}$, 2

H
Hilbert manifold, 140
Hilbert sphere, 6, 121
Homogeneous space, 151
Homomorphism, 88
Hyperbolic spaces, $\mathcal{H}_n$, 5, 149
Hyperboloid hypersurface, 154

I
Injectivity radius, 69
Interpolation, 1
Isotropy group, 68

K
Kendall space, 76
  Kendall shape space, 137
Koszul formula, 90

L
Least-squares, 4
Levi-Civita connection, 5, 90
Locally symmetric space, 100, 116
Logarithm map, $\mathbb{S}^n$, 31
Logarithm map, $\mathcal{P}_n^+$, 151
Logarithm map, $SO(n)$, 52
Logarithm map, $St(n, p)$, 69
Logarithmic map, $\mathcal{P}_+(I)$, 69
Lorentzian inner product, 154

M
Machine learning, 1
Machine vision, 50
Manifold, 2
Mean squared acceleration, 10

Mean squared velocity, 10
Medical imaging, 1
Minimal geodesics, 5, 85
Motion planning, 27

N
Nanomanufacturing, 1
Non-Euclidean spaces, 3
Non-linear spaces, 1

O
Optimization, 4
Orthogonal Lie group, 67
Orthogonality constraints, 5

P
Parallel transport, 5
Parallel transport, $\mathbb{S}^n$, 31
Parallel transport, $\mathcal{P}_+(I)$, 101
Piecewise polynomial, 11
Piecewise smooth curves, 28
  piecewise smooth geodesics, 28
Poincaré ball, 154
Polynomial functions, 3
Prediction, 2
Probability Density Functions, $\mathcal{P}$, 2

Q
q-function representation, 139
Quotient manifold, 67

R
Regression, 2
Riemannian exponential map, 5
Riemannian isometry, 119
Riemannian manifold, 28
Riemannian mean, 28
Riemannian metric, 3, 5
Riemannian metric, $\mathbb{S}^n$, 30
Riemannian polynomials, 3
Rotation group $SO(3)$, 50

S
Second covariant derivative, 4
Shape analysis, 1
Shape space
  shape space of curves, 139
Smooth curve, 2

Space of probability measures $\mathcal{P}_+$, 5, 86
Special orthogonal group $SO(n)$, 2
Sphere $\mathbb{S}^n$, 6, 27
Spherical splines, 27
Square-root density function, 118
Square-root representation, 117
Squared norm, 3
Stiefel manifold $St(n, p)$, 2
Symmetric and positive-definite matrices, $\mathcal{P}_+^n$, 1
Symmetric Riemannian manifold, 50

T
Tangent space, 30
Tangent vector, 30, 51

V
Variational problem, 11
Vector field, 91
Velocity, 4

SPRINGER NATURE

GPSR Compliance

The European Union's (EU) General Product Safety Regulation (GPSR) is a set of rules that requires consumer products to be safe and our obligations to ensure this.

If you have any concerns about our products, you can contact us on ProductSafety@springernature.com

In case Publisher is established outside the EU, the EU authorized representative is:

Springer Nature Customer Service Center GmbH
Europaplatz 3
69115 Heidelberg, Germany

The manufacturer's authorised representative in the EU is Springer Nature Customer Service Centre GmbH, Europaplatz 3, 69115 Heidelberg, Germany. If you have any concerns regarding our products, please contact ProductSafety@springernature.com

Printed and bound by CPI Group (UK) Ltd, Croydon, CR0 4YY
25/03/2026
02078171-0003